Deep Time

Henry Gee is a Senior Editor at *Nature*. He holds a PhD from Cambridge in Zoology and has previously been Regent's Professor at the University of California, Los Angeles. He also contributes to *Le Monde*, *El Pais* and *Die Zeit* and has previously written *Before the Backbone: Views on the Origin of the Vertebrates* (Kluwer, 1996).

HENRY GEE

Deep Time

CLADISTICS, THE REVOLUTION IN EVOLUTION

FOURTH ESTATE • *London*

This paperback edition first published in 2001
First published in Great Britain in 2000 by
Fourth Estate
A Division of HarperCollins*Publishers*
77–85 Fulham Palace Road,
London W6 8JB
www.4thestate.co.uk

10 9 8 7 6 5 4 3 2 1

A catalogue record for this book is available from the British Library.

ISBN 978-0-00-729154-0

Typeset by MATS, Southend on Sea, Essex
Printed in Great Britain by Clays Ltd, St Ives plc

For John and Sue

Contents

List of Illustrations   ix

Introduction   I

I   Nothing Beside Remains   II

2   Hunting Unicorns   45

3   There Are More Things   84

4   Darwin and His Precursors   110

5   The Gang of Four   136

6   The Being and Becoming of Birds   169

7   Are We Not Men?   199

Notes   228
Acknowledgements   252
Index   254

Illustrations

All diagrams and pictures drawn by Majo Xeridat, unless specified below.

1 A chart of Deep Time, compiled by the author from various sources   *4*

2 How the ancestries of Fred and myself are linked, through Deep Time, with our common ancestor at the 'node'   *36*

3 A diagram showing the inferred relationships between myself and my cats, Fred and Marmite   *38*

4 An alternative relationship between myself, Fred and Marmite   *39*

5 Adding the perspective of an outgroup (a pigeon, in this case) allows the resolution of two equally parsimonious cladograms   *42*

6 A cladogram with a fossil in it   *43*

7 'I will be the *First* on Land.' From *Cartoon History of the Universe* by Larry Gonick (Rip-Off Press, San Francisco)   *45*

8 A cladogram showing the pattern of relationships between tetrapods and other lobe-finned fishes. Redrawn by Majo Xeridat from Ahlberg and Johanson's article 'Osteolepiforms and the ancestry of tetrapods', *Nature* vol. 395 (1998), pp. 792-4   *62*

9 Spandrels arise as inevitable consequences of supporting a dome on arches   *100*

10 A cladogram showing the position of *Triceratops* in the pattern of life   *104*

11 The problem of the Salmon, the Lungfish and the Cow   *142*

12 Cladograms and lineages. How one cladogram summarises two or more possible scenarios of ancestry and descent   *146*

13 A cladogram of hominid interrelationships   *209*

Introduction

To claim that it is true is nowadays the convention of every made-up story. Mine, however, *is* true.

Jorge Luis Borges, *The Book of Sand*

How did the dinosaurs become extinct? Why did fishes evolve legs and learn to walk on land? How did birds become airborne? Is humanity the culmination of evolution? Apart from their perennial fascination, questions like these – evolutionary questions, about 'how' and 'why' – are united in implying a narrative in which causes and effects are linked: dinosaurs became extinct *because* of the after-effects of an asteroid impact, or *because* they were rendered obsolete by mammals. Chains of cause and effect may also be animated by purpose: fishes evolved legs *for* walking on land; birds evolved feathers *for* flight; human beings evolved from ape-like ancestors *because* they had bigger brains, could make tools and use language.

Popular views of science assume that cause, effect and purpose can be easily discerned: hardly a day goes by without reports of discoveries of genes *for* homosexuality, good motherhood, breast cancer or alcoholism; of the bones of ancestors and 'missing links'; of explanations of why the elephant evolved its trunk; or warnings that a certain food item *causes* a particular disease.[1] Many of the assumptions we make about evolution, especially concerning the history of life as understood from the fossil record, are, however, baseless.

The reason for this lies in the scale of geological time that scientists deal with, which is so vast that it defies narrative. Fossils, such as the fossils of creatures we hail as our ancestors, constitute primary evidence for the history of life, but each fossil is an infinitesimal dot, lost in a fathomless sea of time, whose relationship with other fossils and organisms living in the present day is obscure. Any story we tell against the compass of geological time which links these fossils in sequences of cause and effect – or ancestry and descent – is, therefore, only ours to make. We *invent* these stories, after the fact, to justify the history of life according to our own prejudices. Nobody will ever know what caused the extinction of the dinosaurs, because we weren't there to watch it happen. All we have are two isolated observations: the apparent absence of dinosaurs since 65 million years ago, and evidence for a catastrophic phenomenon, such as the impact of an asteroid, at around the same time. There can be no certain link between the two. Geological time admits no narrative in which causes can be indubitably linked with effects.

Fossils are never found with labels or certificates of authenticity. You can never *know* that the fossil bone you might dig up in Africa belonged to your direct ancestor, or anyone else's. The attribution of ancestry does not come from the fossil; it can come only from us. Fossils are mute: their silence gives us unlimited licence to tell their stories for them, which usually take the form of chains of ancestry and descent. These stories are like history, of events leading to other events; of succession and defeats; change and stability. Such tales are sustained more in our minds than in reality, and are informed and conditioned by our own prejudices – which will tell us not what really happened, but what we *think* happened. If there are 'missing links', only our imaginations can reconstruct them.

John McPhee, an eloquent writer on geology, coined the term 'Deep Time' to distinguish geological time from the scale of time that governs our everyday lives.[2] McPhee meant the term to refer to the immense intervals, measured in millions of years, discussed as if they were days or weeks in the conversation of geologists; yet, in reality, the intervals of geological time are too long to be readily comprehensible to minds used to thinking in terms of days, weeks and years – decades, at most.

Books on the history of life usually start with an appreciation of the vastness of Deep Time, and try to use some analogy to make it comprehensible in everyday terms. Walter Alvarez, for example, suggests thinking of an interval of a million years as if it were a kind of geological 'year'.[3] On this scale, the dinosaurs became extinct 65 years ago – a lifetime away, just before the outbreak of World War II. Fishes evolved legs and clambered ashore about 360 years ago, in the 1640s, around the time of the English Civil War, and animals with backbones appeared approximately 500 years ago, when the first conquistadors made landfall in the Americas. The first signs of life on Earth appeared more than 3,600 years ago, when Babylon was a major city, and the Earth itself formed 4,500 years ago, around the time that the mythical hero Gilgamesh is supposed to have gone about his heroic business. On this scale, our own species, *Homo sapiens*, is a very recent arrival, appearing only a year ago.

At this point, popular books present a table outlining the formal divisions of geological time and calibrating their passage in millions of years from the present day. My own contribution can be found in Figure 1. I present it because I mention technical names for several geological intervals throughout this book, and you might wish to refer to it.

After presenting a chart of geological time and impressing you with its scale, canonical accounts of the history of life consider that their duty towards Deep Time has been fulfilled, and will move on to telling the story of life as if it were *Hamlet*, a drama that can be understood in human terms. We will learn of the origin and evolution of life, the rise of the fishes and the fall of the dinosaurs, the evolution of the birds and mammals, the primates, and finally of Man. Each species will make its entry and exit like an actor on a stage, with Deep Time as the backdrop.

But apart from telling you that Deep Time is long, conventional accounts never consider the implications of the scale of Deep Time for the way we think about evolution. If, as McPhee says, Deep Time implies intervals more or less incomprehensible to humans, we are entitled to ask whether it is valid to tell stories about evolution according to the conventions of narrative or drama. If it is not, then every story we tell in which causes are linked with effects, and ancestors are linked with descendants, becomes questionable: we can no longer use Deep Time as a backdrop for the stories we tell ourselves about evolution, and how and why we came to be who we are.

Age (millions of years)	Period	Era	Eon
2	Quaternary	Cenozoic	Phanerozoic
65	Tertiary		
144	Cretaceous	Mesozoic	
213	Jurassic		
248	Triassic		
286	Permian	Palaeozoic	
360	Carboniferous		
408	Devonian		
438	Silurian		
505	Ordovician		
543	Cambrian		
1000		Neoproterozoic	Proterozoic
1600		Mesoproterozoic	
2500		Palaeoproterozoic	
4000			Archaean
~4500 (origin of the Earth)			Priscoan

Figure 1. A chart of Deep Time.

Once we realize that Deep Time can never support narratives of evolution, we are forced to accept that virtually everything we thought we knew about evolution is wrong. It is wrong because we want to think of the history of life as a story, but that is precisely what we cannot do. This tension – between Deep Time and the everyday scale of time – is the theme of this book. What we need is an antidote to the historical approach to the history of life; a kind of 'anti-history' that recognises the special properties of Deep Time.

INTRODUCTION

If we can never know for certain that any fossil we unearth is our direct ancestor, it is similarly invalid to pluck a string of fossils from Deep Time, arrange these fossils in chronological order, and assert that this arrangement represents a sequence of evolutionary ancestry and descent. As Stephen Jay Gould[4] has demonstrated, such misleading tales are part of popular iconography: everyone has seen pictures in which a sequence of fossil hominids – members of the human family of species – are arranged in an orderly procession from primitive forms up to modern Man. To complicate matters further, such sequences are justified after the fact by tales of inevitable, progressive improvement. For example, the evolution of Man is said to have been driven by improvements in posture, brain size, and the coordination between hand and eye, which led to technological achievements such as fire, the manufacture of tools and the use of language. But such scenarios are subjective. They can never be tested by experiment, so they are unscientific. They rely for their acceptability not on scientific test, but on assertion and the authority of their presentation.

Given the ubiquitous chatter of journalists and headline writers about the search for ancestors, and the discovery of missing links, it may come as a surprise to learn that most professional palaeontologists do not think of the history of life in terms of scenarios or narratives, and that they rejected the story-telling mode of evolutionary history as unscientific more than thirty years ago. Behind the scenes, in museums and universities, a quiet revolution has taken place.

The architects of this revolution sought ways to discover the pattern of the history of life that are free from subjective, untestable stories. If it is fair to assume that all life on Earth shares a common evolutionary origin, it follows that every organism that ever existed *must* be related to every other. We are all cousins. Every goanna and gourami is a cousin of every gecko and ginkgo that has ever lived, or will live in the future. This must be true, even though we can neither tell who was whose direct ancestor, nor justify any scenarios to support assertions about ancestry and descent.

Before we can understand the history of life, we need to find the order in which we are all cousins, the topology or branching order of the tree of life. This can be done without having to make any prior assumptions about cause and effect, or ancestry and descent. These branching diagrams, which look, misleadingly, like genealogies, are proper scientific hypotheses that can be tested by examining the strength or likelihood of alternative orders of branching – different orders of cousinhood – in the light of the anatomy of the organisms in whose relationships we are interested. As long ago as 1950, a German entomologist called Willi Hennig used these simple principles as a basis for a new way of looking at the living world: Hennig sought to understand creatures in terms of how they shared characteristics with one another, independently of time, rather than in terms of their histories of ancestry and descent. Hennig called his philosophy 'phylogenetic systematics', but it came to be known as 'cladistics' and its practitioners, inevitably, as 'cladists'. The branching diagrams cladists drew up to represent orders of cousin-hood between organisms – patterns of relationship – became known as 'cladograms'.

Cladistics looks only at the pattern of the history of life, free from assumptions about the process of the unfolding of history. It resolves the conundrum of trying to comprehend Deep Time in terms of an unfolding drama. Because of this, cladistics is the best philosophy for the scientific understanding of the history of life as we unearth it from Deep Time. More than a set of techniques, but less than a science in its own right, cladistics is a way of 'seeing', of looking at the products of evolution as they are, and not as we would like them to be.

An important aspect of cladistics, as in all science, is testability. In cladistics, you are asked to find the most likely way in which a set of organisms is related to one another; to estimate, in other words, their relative degrees of cousinhood. This is done with a cladogram. As I show in Chapter 1, if you have more than two organisms, you will find that there is more than one way of drawing a cladogram that links them up. What this means is choice; if there is more than one way in which organisms can be cousins, you are forced to consider the alternatives, and you must find a way of evaluating them all.

How can these alternatives be evaluated? The central test is as old as science, older: Occam's Razor, or the Principle of Parsimony.

That is, if you have to make a choice between explanations, you should choose the simplest. The simplest or most 'parsimonious' cladogram is the one that assumes the smallest amount of evolutionary change. I will discuss what that means for our understanding of evolution later in this book.

Conventional stories about evolution, about 'missing links', are not in themselves testable, because there is only one possible course of events – the one implied by the story. If your story is about how a group of fishes crawled onto land and evolved legs, you are forced to see this as a once-only event, because that's the way the story goes. You can either subscribe to the story, or not; there are no alternatives. In cladistics, you can put the story to the test by studying the fossils and asking yourself how they might have been related to one another. A fossil fish you happen to find *might* have been your direct ancestor, but you will never be able to establish the case, one way or the other.

Such study might turn up some surprises. For example, your evidence might show that more than one group of fishes crawled ashore independently; or that other groups of fishes evolved legs, but stayed underwater. By offering choices, cladistics opens our eyes to surprising new discoveries and possibilities of evolution otherwise hidden from us. Conventional wisdom has fishes crawling ashore, evolving limbs and progressing, inexorably, to amphibians, to reptiles, to mammals, and to Man. Cladistics suggests that things might have been otherwise.

Testability is a central feature of the activity we call science. Some have sought a kind of special dispensation for palaeontology as a 'historical' science, that it be admitted to the high table of science even though palaeontologists cannot, classically, do the kinds of experiments other scientists take for granted. You cannot go back in time to watch the dinosaurs become extinct, or fishes crawl from the slime to become amphibians. More pointedly, you cannot, as Stephen Jay Gould discussed in his book *Wonderful Life*, go back in time to see what other things might have happened instead, had circumstances been slightly different. What if the asteroid had missed the Earth, sparing the dinosaurs? What would have happened if the fishes decided to stay underwater after all? In either case, would we be here? We cannot see what nature would have done had she been able to rerun the tape of evolution. Were we able to witness such a rerun, would the outcome have been different from what we see, as Gould argues, or very much the same?[5] In strict, scientific terms, such questions are meaningless. The problem is that what we see before us is the result of a once-only experiment in history. Because it happened only once, it is not accessible to the reproducibility scientists usually require. This is not possible in palaeontology except in our imaginations.

However, palaeontology either is, or is not, scientific and you may ask whether the particular problems that palaeontology has with its subject – Deep Time – should be allowed to mitigate its inability to reproduce experiments in the approved scientific manner. It should not. To see palaeontology as in any way 'historical' is a mistake in that it assumes that untestable stories have scientific value. But we already know that Deep Time does not support statements based on connected narrative, so to claim that palaeontology can be seen as a historical science is meaningless; if the dictates of Deep Time are followed, no science can ever be historical.

Palaeontology read as history is additionally unscientific because, without testable hypotheses, its statements rely for their justification on authority, as if its practitioners had privileged access to absolute truth – 'truth which can be known', in the words of the late palaeontologist and cladist Donn Rosen (whose views I discuss in Chapter 5). Whether you believe the conventional wisdom that, for example, our own species *Homo sapiens* descended in seamless continuity from the pre-existing species, *Homo erectus*, depends not on the evidence, for the fossil evidence is mute, but on whether the *presentation* of the evidence conforms to your prejudices, or on whether you choose to defer to the authority of the presenter. The assumption of authority is profoundly, mischievously and danger-ously unscientific. It conflicts with how we are taught science from our earliest years: that the scientific method should be rigorously democratic, that statements from authorities in a field should be as subject to scrutiny as those emanating from the most humble sources, even a novice. Nobody should be afraid to ask a silly question.[6]

Cladistics has remained true to this simple view of science. When, as a student, I spent a summer working on fossil fishes at the Natural History Museum in London, my innocent views were accorded the same respect as if they had come from the head of the Fossil Fish section, the late Colin Patterson, one of the world's most respected workers on fossil fishes, and one of the most influential advocates for cladistics. As I discuss in Chapter 5, Patterson wished to replace the elitist, authoritarian presentation of old-fashioned museum displays with an approach to science that encouraged the participation of a wider audience, represented by the museum-going public. My opinion received equal value because Patterson, for all his eminence, had no special access to the truth that would be denied to me.

The story of human interaction with fossils represents an example of how experience and belief have a powerful effect on interpretation, and demonstrates why scientific truths can only ever be temporary. Today, we see fossils as the remains of creatures which once lived. However, this nature is not inherent in the fossils. It is our immersion in a century and a half of Darwinian thought, not the fossils themselves, that gives us the capacity to see fossils as kin to things which were once as alive as you or me. If this were not the case, we would interpret fossils differently. In societies without science in the sense that we understand it today fossils were seen as signs of divine or diabolical action; as the bones of giants or the teeth of dragons; or as the bones of creatures which perished in Noah's Flood. Observers innocent of science, ignorant of religious or cultural tradition and incapable of imagination would no doubt see fossils only as rocks.[7] All these views could be as valid to their respective beholders, as is the apprehension of fossils as the remains of living organisms to us. We cannot be certain, therefore, that our current understanding of fossils is not as provisional as these earlier ones.

This line of reasoning appeals to contemporary philosophers and critics who regard with detachment the conventional claims of scientists to be rolling back the frontiers of ignorance, unveiling universal truths and so on. Such critics suggest, instead, that what scientists find out is conditioned at least as much by their cultural heritage as by objective reality. It is hard to argue with this point of view, borne out, for example, by the way our approach to fossils has changed over the ages. Looking at science from this perspective, you might ask whether cladistics itself – in its embrace of patterns of relationships rather than lines of ancestry, and its insistence on truths that are relative and provisional rather than absolute and final – falls foul of these contemporary prejudices.

The answer must be 'no'. Cladistics is, in truth, somewhat reactionary in that it embodies a purist attitude to science that seeks to establish traditional, scientific values such as the importance of objectivity, the testability of hypotheses, and the provisional nature of the results of such tests. A cladist might say that the *whole* of science ought to be similarly austere, in that those scientists who claim to unveil universal truths have no business making such claims in the name of science.

Without cladistics, palaeontology is no more of a science than the one which proclaimed that the Earth was 6,000 years old and flat – and *then* had the effrontery to claim Divine sanction for this view.

Nothing Beside Remains

'My name is Ozymandias, king of kings:
Look on my works, ye Mighty, and despair!'
Nothing beside remains. Round the decay
Of that colossal wreck, boundless and bare
The lone and level sands stretch far away.

Percy Bysshe Shelley, *Ozymandias*

I climb to the top of the ridge to get a better look at the view. Mazes of badlands fall away before my gaze; dry gullies opening just a few feet away spread out, broaden and meander into a yellow-green landscape. I am in East Africa, more precisely a part of north-western Kenya, the valley of an ephemeral river called the Topernawi that drains – if such a word applies to a river of sand – into Lake Turkana, at a point some twenty kilometres southeast of where I stand.

Tussocks of grass punctuate the red-grey earth. The ground is patterned by small thorn bushes and dotted with acacias, all with the same flat tops, as if a glass sheet were covering the scene just fifteen feet from the ground. In the distance, the acacias seem to get smaller and closer together as they converge on the dense, grey-green line of the river bank. Further away still, beyond the river, the acacias, now too small to make out individually, thin out into a piebald jumble of hills. Silence surrounds me.

I take a photograph. The result will be disappointing, failing to capture the stillness, the growing heat of the morning, and the cool relief offered by the gentle breeze that met me as I scaled the ridge, as steep and slippery as a pitched roof. After taking the picture, I look down the slope of concrete-hard pebbles, each one a worn, dried fragment of calcareous ooze deposited at the bottom of a river that flowed here around 3.3 million years ago.

Placing one foot squarely on each side of the ridge crest, I notice a glint of white – a fossil bone, weathering out of the soil between my feet. I crouch down to examine the bone more closely. It looks like a real prize, not the knobbly end of a limb bone (for these are common), but the delicate curve of part of a braincase that was perhaps the size of a clenched fist when complete, the sides curving up to a central crest, and showing the thin, wavy lines of sutures. The patch of smooth, almost geometric curvature, no more than an inch square, makes a sharp contrast with the rough texture of the ground.

Taking my penknife, I scrape at the ground around the bone to try to see a little more. I have a hunch that the small curve of bone before me extends further into the ground, arcing tightly to define the back end of a roughly egg-shaped mass. My imagination, too quick for my mind to rein in, makes a skull before my eyes. The front end of the fossil looks like the twin scrolls of a dog-like snout. The space between the front and the curved back end – the part I first spotted – is mangled beyond recognition. With a bit of effort, hope and imagination, I might just make out the shape of an eye socket.

Or perhaps my imagination is running away with me. Even if this specimen really is a fossil skull, I have no idea what it might have belonged to. I run through a number of possibilities in my mind's eye, trying to match each mental image with the specimen on the ground. My first thought is that it might have been the skull of a turtle. I leave the specimen for a few minutes so I can prospect down the slope. Perhaps some other bones might have been attached to the fossil but had weathered out and fallen down the hill. A minute's work turns up a few pieces of fossil turtle shell. This is relatively common here: so common that the distinctive, trapezoidal, heavily sculpted plates are rarely collected. However, I wonder whether they might produce clues to help me identify my mysterious find. In the end, I decide that I cannot make progress without expert help. I had done my PhD thesis on fossil cattle bones, not turtles. Had I found a few antelope limb bones, I could have told the difference between tibia and femur, humerus and radius, and even have been able to tell whether it came from the left leg or the right. But turtles do not figure in my mental repertoire. I mark the spot with a cairn of rocks, protecting the fragile fossil against further weathering, and the hooves of wandering Turkana goats.

That evening I tell Meave Leakey of my puzzling find. Meave is the head of the Department of Palaeontology at the National Museums of Kenya in Nairobi. She promises to take a look at the mystery skull next day. I am a guest on her summer expedition to survey and collect from the fossil beds on the western shore of Lake Turkana. The next day, Meave scales the ridge with me and Eleanor Weston, an English post-doctoral researcher who is an expert on fossil hippopotami.

Crouching on the ridge over the emerging skull, Meave does some thinking aloud, as she runs through her own mental zoo. The specimen is not a turtle, nor any kind of reptile. It is a mammal, possibly a carnivore – a member of the mammalian order that includes cats and dogs, otters and weasels, bears and lions – but then again, it could be nothing. Eleanor collects the loose chips of fragments into plastic bags, as Meave paints the exposed parts of the specimen with a rapid-hardening synthetic resin, to protect it from further damage, and we leave it to dry for another day.

This time we completely surround the fossil with rocks, roofing it over with a large slab, as if it were a Gallic chief under his dolmen. This is done to protect the fossil from hyenas, which cannot resist the smell of the hardening agent. Fossil-hunting is full of stories of the Ones That Got Away, of marvellous fossils which, when coated with resin and left exposed, had vanished by the following day, carried off by these nocturnal thieves.

The following day, Meave leaves me atop the ridge with some old dental picks to see how far I can get excavating the fossil from the ground. Old dental picks are the Swiss-Army knives of palaeontology, wonderful tools for wheedling fossils out of the ground, by removing the surrounding rock, grain by grain. The hardness of the rock defeats me. I lack the confidence of an experienced fossil preparator. Consequently, I'm always afraid that a little too much pressure applied to the pick might cause it to skid off the hard rock and damage the softer, more delicate bone. I poke prissily around the specimen and don't get anywhere. Excavating fragile fossils from hard rock is a skill that takes time to master. Onyango, one of

Meave's technicians, relieves me at my post. He teases the fist-sized lump from the ground without drama or fuss, and carries it back to camp on its own bed of sand, in a round-bottomed steel bowl, or *korai*. From there, the fossil is taken back to base camp on the shore of Lake Turkana, and thence to the Museum in Nairobi to be further investigated, its identity still uncertain, though opinion is converging on its being a carnivore.

It looks rather like the skull of the modern African civet cat (*Civettictus civetta*), a relative of the mongoose, but much larger. Tantalisingly, the skull could have belonged to the giant civet (*Pseudocivetta ingens*), an extinct relative of the African civet. Bones and teeth from this rare creature have been found at Olduvai Gorge in Tanzania, and at Koobi Fora, on the eastern shore of Lake Turkana. If the skull really did once belong to a giant civet, it would be the first example ever found on the west side of Lake Turkana and, at 3.3 million years old, the oldest known example from anywhere by a million years; I would have held in my hands the only evidence of hundreds of thousands of generations of giant civets that must have existed between this example, and the fossil closest to it in time. When my giant civet was alive, the next animal in its species to leave any trace of itself to posterity to date lived a million years in the future.

At the time, we were working a square kilometre of jumbled badlands, near where the Topernawi meets another sand-river, the Lomekwi. In the 1980s, Meave and her colleagues, John Harris from the Los Angeles County Museum of Natural History, and geologist Frank Brown from the University of Utah, surveyed dozens of potential fossil-collection sites for a hundred kilometres up and down the western shore of Lake Turkana.[8] Some were deemed worthy of more intensive investigation: this site, a confused and mazy terrain between the two sand-rivers, was named 'LO5' (site number 5 on the river Lomekwi), and was marked as one of the more promising sites.

LO5 lives up to its promise. The members of the crew are finding fossils on the surface wherever they look. Each time someone finds a fossil, they mark its place with a small cairn. Eleanor and Meave trail from cairn to cairn, collecting these fossils, identifying them where possible, noting something of their location, and entering the details in a notebook. A few are like the tips of icebergs, marking places where a skeleton has begun to weather out of the ground. If this happens, the place is marked down for concentrated surveying and digging. My own maybe-civet was found not far from a place where some of the team had noticed a concentration of monkey bones leaching from a hillside. A dig reveals a near-complete skeleton of an extinct baboon, buried a foot underground, halfway up a hillside. Meave is very pleased, as this may be the most complete specimen of this species of monkey ever found.

Each fossil that Meave and Eleanor collect receives a field number of a special kind. Each number starts with the prefix KNM-WT, for 'Kenya National Museums, West Turkana', followed by the number.

Some of the world's most famous fossils have KNM-WT numbers. They include KNM-WT 15000, a near-complete skeleton of a young male *Homo erectus*, found buried under a tree-root at a place called Nariokotome,[9] some way to the north of LO5. A few days before finding my mystery skull, Meave and Frank Brown (who had flown in from Utah to map the area's geology) took me to the site where, in the 1980s, the Leakeys and their colleagues had found KNM-WT 17000, the so-called 'Black Skull', an impressive specimen of *Paranthropus aethiopicus*.[10] *Homo erectus* and *Paranthropus aethiopicus* are both hominids. That is, they belong to the family Hominidae. All the species of hominid that ever lived are now extinct, except one, *Homo sapiens*, the species that includes every human being now alive.

Meave and Eleanor, sitting under an acacia in the camp next to the Topernawi, give my mystery skull its own 'KNM-WT' number. It is a modest fossil, but it has illustrious company.

How is it that East Africa comes to have such a rich fossil heritage, especially one that sheds light on human prehistory? Why are we looking for fossils just here, on the shores of Lake Turkana, when we could be looking for them in more comfortable places closer to home, where there are roads and telephones and not so many snakes and scorpions?

Fossils are where you find them, not where you want them. They are not found spread uniformly through the Earth's sediments. They are granted to posterity only thanks to accidents of geology. East Africa has the greatest accident of them all – the Great Rift – which just happens to be wealthiest in fossils of the right geological age to encompass the origin and early evolution of the hominids. As a bonus, this region also happens, by coincidence, to be the area where humanity is believed to have originated.

The Great Rift is an ocean in the act of being born. Deep under Africa, something has been stirring. Forces beneath the Earth's crust, forces we do not fully understand, keep the molten rock on which East Africa rests hotter and more agitated than the rocks beneath neighbouring landmasses and oceans. The stream of molten rock, welling upwards from the Earth's interior, meets the hard barrier of Africa's undersurface and can go no further. Thus impeded, it splits in two, one tributary travelling westwards beneath Africa's roots, the other towards the east, beneath the Indian Ocean. As the two streams of hot, viscous rock separate, they drag and snag on Africa above, tearing it in two. Lands that were once neighbours become separated by a slowly widening valley. This is the Great Rift.

The Rift extends for more than four thousand miles, from the Dead Sea in the north, down the Red Sea, into the Danakil Depression of Eritrea, into Ethiopia, and southwards into East Africa as far as Mozambique. The land on either side often rises, like a giant welt around a cut, to commanding heights, before plunging, in dizzy escarpments, to a valley floor as much as 9,000 feet below. Vast amounts of sediment, weathered from the escarpments, slump into the growing chasm, forming great terraces and ramparts. The strains in the Earth's crust, set up by the rifting, occasionally punch through the sediments to the surface to create volcanoes such as Kilimanjaro.

Water, finding its own level, collects in the Rift, creating a chain of lakes: Victoria, Tanganyika, Malawi, and many others less familiar: Turkana, Chew Bahir, Kivu, Naivasha, Baringo. With rivers and lakes come sediments borne by the action of water, which, turning to mudstones and sandstones, preserve a small sample of the fossil remains of the wildlife that lived along the river banks and by the lakes.

The slow widening of the Rift has something of the irresistible force exerted by the growth of the roots of a tree, a force that seems hardly measurable on the scale of the everyday, yet given enough time, it can split walls and fell buildings. The roots of a tree may do their work in fifty years and pull down a house, but the Great Rift has taken, so far, thirty million years to unfold. By chance, this interval encompasses the period in which hominids emerged as a distinct group of creatures.

The earliest known hominid is *Ardipithecus ramidus*, whose remains were buried in Ethiopia 4.4 million years ago.[11] A little over four million years ago, another species, *Australopithecus anamensis*, lived in the Turkana basin. Meave and her colleagues found its remains to the south and east of Lake Turkana.[12] The list of famous hominid fossils goes on: it includes *Australopithecus afarensis*, better known to the world as 'Lucy'; *Paranthropus boisei*, unveiled as 'Nutcracker Man' or 'Zinj'; *Homo habilis*, more familiarly 'Handy man' – all come from the Rift.[13]

Lake Turkana extends like a bright green jalapeño pepper southwards from Ethiopia into the scrub desert of north-western Kenya. It is filled by the southward-flowing river Omo that drains the Ethiopian Highlands. As it descends into the Rift, the river wears a scar through rocks that were deposited as sediment as long as five million years ago, so that the Rift has split open like the pages of a book.

Lake Turkana itself covers almost two and a half thousand square miles. Of all the Rift Valley lakes, only Victoria, Tanganyika and Malawi are more extensive. A feature of the Earth's physical geography as immense as Lake Turkana looks as though it should have been there for all eternity. But the lives of lakes are ephemeral when measured against the slowness of the Rift's formation. Over millions of years, lakes in the Rift have repeatedly spread and shrunk, moving to the dance of the shifting strata below. Four million years ago, Lake Turkana covered three times its present area. Yet for 85 per cent of the period between then and now, the lake hardly existed. Given the perspective of millions of years, we are lucky to find a lake here at all. Through all this time of change, the landscape of the Rift has been remarkably constant. A visitor to northern Kenya three million years ago would have seen more or less the same scrub desert that first greeted me.

Given enough time, great changes will be evident, even here. For the next few tens of millions of years, the Rift will continue to widen as the two parts of Africa are pulled apart. There will come a time when the land within the Rift will slump so far that it will be below sea level. This has already happened in the Danakil Depression in Eritrea, and in the Jordan Valley, at the northern end of the Rift. The Red Sea is an arm of the Indian Ocean that has already invaded a part of the Rift. Eventually, the valley now occupied by Lake Turkana will also be occupied by an arm of the sea. The day will come, millions of years from now, when East Africa will be entirely surrounded by water. It will be a new island continent. Were we to stand on the western shore of Lake Turkana in thirty million years time, we would no longer be able to see to the other side. We would be standing on the beach of a new ocean.

Like the view, the palaeontologists of the future will also have changed. If they are our descendants, they might not look like us at all. Alternatively, these future palaeontologists could have had some completely different heritage, closer to rats, whales or cockroaches than to humans. Much can happen in thirty million years. Mammal species are thought to last, on average, only a few million years before they become extinct or evolve into something else. So, even if the palaeontologists of the future are the descendants of humanity, the human species as we know it will almost certainly have become extinct by then: as extinct as, say, *Paranthropus aethiopicus* is today. If we are lucky, our own remains will be excavated with expert care, and put on display in museums to be admired by visitors of unguessable countenance.

The search for human origins in the Rift is synonymous with the name of Leakey.[14] Louis Leakey was the son of a missionary who came to preach the gospel to the Kikuyu tribe in 1902. Louis dug up fossils all over what is now Kenya and northern Tanzania. In 1948, Louis's wife Mary discovered the first skull of the fossil ape *Proconsul africanus* on Rusinga Island in Lake Victoria. Louis and Mary spent three decades, on and off, working at Olduvai Gorge before they discovered their first hominid there in 1959. This was

'Zinj', now known as *Paranthropus boisei*. Mary spent most of the remainder of her life documenting the stone-age archaeology of Olduvai Gorge. Richard, one of their three sons, was brought up with fossils, and was drawn to them despite early reluctance. In 1972, when Richard's colleague Bernard Ngeneo discovered a skull of *Homo habilis* at Koobi Fora, on the eastern shore of Lake Turkana, it was Richard's wife Meave who assembled the cranium from a box of fragments. Meave and Richard's two daughters, like their father, spent their summer holidays in field camps, and now one of them, Louise, is running her own field programme.

The Leakeys owe many of their discoveries to their elite corps of hominid-finders, the 'Hominid Gang'. Most of the group that found fossils with Richard Leakey at Koobi Fora in the 1970s have retired. One, Peter Nzube, has come out of retirement for the 1998 field season. Nzube can probably claim to have collected more hominid specimens than anyone else alive – and to have thrown away twice as many pebbles which, when first seen against the ground, looked so promising. The work is hard, hot and boring.

Nzube would not have got through forty years of hard work without a sense of humour, particularly when asked stupid questions by first-time visitors. I had picked up a pebble and thought it looked something like a hominid tooth, albeit very worn, and only if one had an active imagination. Nzube happened to be walking near.

'Nzube', I asked, 'what's this?'

Nzube speaks little English, and what he has, he uses with economy. He took the pebble, smiled, and said 'stone'. Gentle laugh-lines creased his face as he handed the pebble back to me as carefully as if it had been the discovery of the decade.

The retirement of the Hominid Gang means that Meave and her colleagues at the National Museums of Kenya are recruiting and training the next generation of hominid finders. For the most part they are young Turkana men in their late teens or early twenties, born and raised in this area. Some come from the market town of Lodwar, several hours' drive away over dirt roads. Others started out as boy goatherds, living in palm-leaf huts in small thornbush enclosures, part of a pastoralist heritage.

Some have managed to finish high school and speak fluent English. Others never get as far as finishing. In this remote, rural part of Africa, a disaster such as inter-tribal conflict, or just bloody banditry, is often likely to strike before the older children can complete their schooling, and they are compelled to leave and earn a living to support their families. Being a fossil-hunter and camp worker earns ready money for two or three months.

The week before I arrived, some weary and ragged-looking Turkana passed through the beach camp, from the north. They were refugees from a stock raid, perhaps remnants of families massacred in the casual genocide of the region. 'I am going south', said one, with great dignity, 'to start a new life.'

Robert is one of the new breed of hominid-hunters. He is a Turkana, a happy-go-lucky former goatherd, never without a wave and a smile. When, as an adult, he first went to Nairobi, somebody had to hold his hand each time he crossed the street, or he would have been mown down by the dense traffic alien to him. In the country of the Turkana, made-up roads and motor vehicles are rare. Nevertheless, he has acquired an eye for fossils and the respect of the older, more experienced fossil-hunters. The day before I arrived in the camp, Robert found a scatter of hominid teeth on a raised patch of flat ground, about the size of a tennis court, not far from where I was to find my mystery skull a week later. Robert's discoveries were rather different from an orderly array of molars freshly pulled in the dental surgery. They were small, cracked, broken and blackened by time. Together, they looked like a broken string of black pearls almost indiscernible from the background of black, grey and brown pebbles from which Robert's keen eye had plucked them.

Not surprisingly, Robert was pleased with his finds. But they were not to be his last. The following day, a short distance away on a stony bank beneath a thorn tree, Robert found two hominid incisors together, side by side in a fragment of jawbone. He smiled as he held these conjoined front teeth in his palm. The teeth were those of a child who lived and died here a little over three million years ago.

Others on the team also found hominids. Gabriel is a young Turkana man from Lodwar, slightly older than Robert, and less happy-go-lucky. He is less inclined to hang out with the boys; he is more of a loner, a thinker. His ambition is to go to art school and become a painter. But his artist's eye is already fully developed, as he has an uncanny knack for finding hominid fossils. Fossil-finding is largely a matter of luck, but all good palaeontologists know that you can make your own luck by having a good 'search-image' of what you want to find. So, before looking for fossils, Gabriel spends hours studying research papers and high-quality plaster-casts of hominid teeth and jaws found in previous field seasons. Even before he goes out onto the site, his mind is imprinted with the shapes and textures of hominids. While I was at LO5, Gabriel found a hominid specimen on each of four consecutive days. All were tooth fragments, except one: a part of a toe bone no bigger than a grain of rice.

One specimen was revealed when Gabriel and I sieved a pile of sediment from the area where Robert found his string of pearls. When important fossils are found on the surface, the ground is stripped of large rocks and pebbles. The soil, down to a depth of a few centimetres, is removed and sifted in big steel-mesh box sieves with wooden frames, to see what else there is to be found. Two people work a sieve at once, first winnowing away the superficial dust in the wind, then sitting down to pick over every grain individually. When everything has been picked over, the sieve crew shovels another pile of earth into the box and repeats the process. It is hard work, tough on back and knees, but it ensures that as many fragments as possible of a broken bone or tooth are gathered, so they can be glued together later. In this way, we hoped to collect a few more pieces of tooth to add to Robert's initial find.

Gabriel and I are sitting at the sieve while Robert and Onyango clear the ground, collecting more sediment to be sieved later. We pick over large pebbles, throwing them out onto the ground, and scrabble among the smaller pieces like prospectors panning for gold. We chat of this and that, until Gabriel suddenly interrupts his own sentence to home in, with casual confidence, on a sheared half of a hominid molar lying in the litter of pebbles and dust.

It has been a productive week. Nzube, Gabriel, Robert and others on the team have unearthed hominid remains; no complete skulls or skeletons, for these are rare indeed, but evidence enough that could, after study, reveal something about the hominids that lived in this region 3.3 million years ago. In all, it has taken approximately 250 man-hours of work to produce enough hominid fragments to half-fill a tin box that Meave carries around on the passenger seat of her truck. Almost all the specimens were pieces of tooth. It does not sound like much, given all that effort, but it is more than most fossil-hunters expect, even from a site that had already yielded a few hominid bones and had earlier been marked as promising.

Before I told everyone else about my own find, straddled on that ridge overlooking an expanse of space and, figuratively, an expanse of time, I wondered fleetingly if it might have been part of a hominid – perhaps half a tooth, like the one Gabriel found. In my mind I was already holding the fragment between finger and thumb, turning it over in the light. The question immediately presented itself: could this fossil have belonged to a creature that was my direct ancestor?

It is possible, of course, that the fossil really did belong to my lineal ancestor. Everybody has an ancestry, after all. Given what the Leakeys and others have found in East Africa, there is good reason to suspect that hominids lived in the Rift before they lived anywhere else in the world, so all modern humans must derive their ancestry, ultimately, from this spot, or somewhere near it. It is therefore reasonable to suppose that we should all be able to trace our ancestries, in a general way, to creatures that lived in the Rift between roughly five and three million years ago.

So much is true, but it is impossible to *know for certain* that the fossil I hold in my hand is my lineal ancestor. Even if it really was my ancestor, I could never know this unless every generation between the fossil and me had preserved some record of its existence and its pedigree. The fossil itself is not accompanied by a helpful label. The truth is that my own particular ancestry – or yours – may never be recovered from the fossil record.

The obstacle to this certain knowledge about lineal ancestry lies in the extreme sparseness of the fossil record. As noted above, if my mystery skull belonged to an extinct giant civet, *Pseudocivetta ingens*, it would be the oldest known record of this species by a million years. This means that no fossils have been found that record the existence of this species for that entire time: and yet the giant civets must have been there all along. Depending on how old giant civets had to be before they could breed (something else we can never establish, because giant civets no longer exist for us to watch their behaviour), perhaps a hundred thousand generations lived and died between the fossil found by me at site LO5 and the next oldest specimen. In addition, we cannot know if the fossil found at LO5 was the lineal ancestor of the specimens found at Olduvai Gorge or Koobi Fora. It might have been, but we can never know this for certain. The intervals of time that separate the fossils are so huge that we cannot say anything definite about their possible connection through ancestry and descent.

In our everyday experience, on the scale of days and years, events can be ordered in time with a fair degree of certainty. We can be confident about what is cause and what is effect, and that one can be linked with another. In our everyday experience, individual events can be linked together to form a continuous narrative. You can see this to be true by looking back at your own working day. Perhaps you can remember four things: rushing to catch the bus in the morning, an important meeting with your boss, having a drink with a colleague after work, and coming home in time to catch the news on TV. These are four separate snapshots from your day, but each is loaded with so many references to the others, and the general context in which all the events take place, that you need never worry that they are not sewn into a narrative, the unfolding story of your life. Your life makes sense through experience; you are surrounded by clocks and calendars, diaries and schedules, familiar things, places and people. The background to your life hardly changes at all; you do not, as a rule, start each day in a different house, commute to a new job by an unfamiliar route, and come home to a spouse and children you've never met. Such sudden changes belong to nightmares. In real life, we are used to continuity. If things change, they tend to change slowly: the car is dirtier than it was yesterday; the grass in the garden has grown a little longer; people grow older.

Our perception of the passage of time hangs on the events within it. Time in this sense is not an independent entity on which events rest, like a conveyor belt transporting objects from the past to the future. As the memory of your own working day tells you, the perception of time as a narrative is a product of the density of events, and their proximity and relatedness to one another in chains of cause and effect. Because events in our everyday experience are profuse and connected, everyday time appears continuous.

Our perception of time changes when events become progressively divorced from the chronological context of our everyday experience. For example, look back at your family photograph albums. Photographs taken in the past few years are full of meaning for you. You can remember who is who in the picture, when and where the picture was taken, and even what was said at the time; your engagement with the picture is enriched by its context, the interconnecting web of events that connect that one snapshot with a whole host of other events, to make a story, part of the ongoing narrative of your life.

Now, look at older photographs, perhaps from that vacation you had a few years ago. The enjoyment of vacations, so vivid at the time, is soon lost when you get back to the humdrum routine of home. The snaps you had developed – you had always meant to label them while the memories were still fresh, but what with one thing and another, you never got round to it. Now, years later, it's too late – you can't quite remember who's who in the pictures, exactly where they were taken, or in which order. Even the year escapes you unless you have some independent means of recalling it; the year that the baby took her first steps, the year we bought the new car, and so on.

Turn the pages, back and further back, until there are no more pages, and all you have is a box of unmounted photographs left to you in the will of a remote great-uncle. None of your own experience allows you even to put these photographs in the correct chronological order. The pictures are of long-dead relations whom you have never met, and of whose existence you were previously unaware.

This happened to me quite recently, after the death of my mother's Uncle Henry, who had been a German émigré. Uncle Henry was born in 1904 and grew up in Berlin in the last days of the Prussian Empire. He came to Britain in 1938, a refugee from the

Nazis. He lived into his eighties, but before he died I remember him telling me how, when he was a truculent four-year-old, his mother and aunt had dragged him to Potsdam to visit the Kaiser's palace. Prussia, the Kaiser, Potsdam – to me, names from history as antique as Ozymandias. It seemed strange, surreal almost, to have them come alive before me, out of the mouth of a relative, then still very much alive, sitting in an ordinary flat in north London.

Going through Uncle Henry's papers, my mother found a picture of a soldier in the uniform of the Kaiser's army. This dated the picture to the time of World War I, and we guessed that the soldier was a relative, given what we knew about Potsdam and Uncle Henry's early life, but there was no writing on the back that might have helped us establish the context. Without the distinctive dress uniform, complete with coal-scuttle helmet, we would have had a hard time even dating it. It could have been a picture of anybody.

When confronted with such an image, no connection of shared events – of narrative – exists between you and the person in the photograph. The event shown in the picture is lost in time, free from any context that might tie it to the present. When the fading of memory dulls the skein of your experience, it becomes difficult to arrange isolated events of your own life, as recorded by family snapshots, in any reliable order. When the life concerned is not your own, but that of another – a person who lived a long time ago in a foreign country – the task becomes impossible, especially when that person is dead and cannot be interviewed.

Looking back further still, we can see that our knowledge of past history – by implication, a narrative – is determined by the density, connectedness and context of events. If events are isolated and disconnected, history breaks down. Students of ancient history, or of more recent intervals such as the Dark Ages in Britain, for example, have problems simply working out the order in which things happened, such as which king reigned before which. These problems are worsened by sparseness of documentation, of context. Between AD 410, when the Romans quit Britain, and AD 597, when Augustine arrived on the coast of Kent, there are few events in British history which can be treated as fact rather than conjecture.[15] This is an interval of time equivalent in length to that between the

Napoleonic Wars and the present day. No wonder, therefore, that this period is full of myth and legend; all sense of connected history and the passage of time ceases, filled instead with the timeless, retrospective romances of King Arthur and the Knights of the Round Table.

The disconnection and isolation of events worsens further as centuries turn into millennia, tens of millennia and finally into millions of years: intervals so vast that they dwarf the events within them. Events become disconnected, separated like stars by gulfs of space measurable not in miles, but in light-years. This is geological time, far beyond everyday human experience.

This is Deep Time.

Deep Time is like an endless, dark corridor, with no landmarks to give it scale. This darkness is occasionally pierced by a shaft of light from an open door. Peering into the lighted room, we see a tableau of unfamiliar characters from the lost past, but we are unable to connect the scene before us with that encountered in any other room in the corridor of time – or with our own time. Deep Time is fragmented, something qualitatively different from the richly interwoven tapestry of time afforded by our everyday experience, what I call 'everyday time' or 'ordinary time'.

A fossil can be thought of as an event in Deep Time. Compared with the immensity of time in which it is found, a fossil is a point in time of zero extent: a fossil either exists or it doesn't. By itself, a fossil is a punctuation mark, an interjection, even an exclamation, but it is not a word, or even a sentence, let alone a whole story. Fossils are the tableaux that are illuminated by the occasional shafts of light that punctuate the corridor of Deep Time. You cannot connect one fossil with another to form a narrative.

So there I was, confronted with a fossil that might have been half a tooth of a hominid, a scrap of flotsam from the ocean of time. Let us give a name – Yorick – to its deceased owner. Yorick *might* have been my lineal ancestor but we can never establish this for certain.

The events of Deep Time – fossils – are so sparse because an animal, once dead, rarely becomes a fossil. A million years passed between one fossil of *Pseudocivetta ingens* and the next. The process of fossilisation and discovery is a concatenation of chance built upon chance. It is amazing that anything ever becomes a fossil at all.

Yorick was just one hominid among many kinds of animal that lived along a tributary stream of the ancient river Omo, which flowed through the present-day Turkana basin, just over three million years ago. The stream was full of fish, turtles and crocodiles. Many other animals made their way down to the stream to drink. There were rhinos, hippos, elephants, monkeys and various kinds of antelope, all animals familiar from the modern African scene. But some of these species were not precisely like the modern forms. Another kind of pig stood in for the extant warthog, for example, and the elephants were of different species from the modern African elephant.

Some of the animals looked very different indeed from anything one might see on safari today. There were deinotheres, animals that looked something like elephants, but larger, with down-turned tusks emerging from the lower jaws rather than the upper; there were sivatheres, distant relatives of the giraffe, but which looked more like elk; and *Homotherium*, a kind of sabre-toothed cat.

The only species of hominid known to have lived in the Rift in Yorick's time was called *Australopithecus afarensis*. This creature probably looked somewhat ape-like, though it walked as erect as any modern human and stood only a metre tall. Its remains have been found up and down the Rift from Ethiopia to Tanzania, and possibly as far west as Lake Chad, halfway across Africa. The famous hominid skeleton named 'Lucy', discovered at Hadar in Ethiopia in 1974 by Donald Johanson and his colleagues, is perhaps the best known example of *Australopithecus afarensis*. Whereas Johanson's team got about forty per cent of a skeleton, we must make do with half a tooth.

Yorick could have been a member of this species, *Australopithecus afarensis*, by default, simply because we know of no other hominids living in East Africa at that time. But there's an old saying in science: that absence of evidence isn't evidence of absence. Yorick could have belonged to some other species entirely. The half-tooth could be the first record of this species – and, so far, the only record that this species ever existed. Imagine – an entire species can evolve, thrive, decline and disappear and leave just half a tooth as a memorial to its existence. You can only wonder how many species have come and gone, leaving no record at all.

Apart from the fact that he had at least half a tooth, we know nothing of how Yorick lived his life, or the manner of his death. Nevertheless, we know from observing animals today that the bodies of dead animals soon disappear. Perhaps Yorick's body – dead but still whole – was carried off, dismembered and consumed by a carnivore, such as a hyena or a jackal. Any bones left over were chewed over and scattered. After a short time, the remaining pieces of Yorick were spread over a wide area. The bones in this residue dried out and cracked in the sun and wind. Yorick's teeth, coated with tough enamel, were more resistant than bone to the kinds of damage suffered by a dead carcass. This is why most of the fossil remnants of mammals are teeth, sometimes attached to fragments of jawbone.

A storm and subsequent torrent washed the little of Yorick that was left – a jawbone bearing several teeth – into the river, where it was buried in sand. These events would have taken place over a few days, a few weeks at most. But the next part of the story took more than three million years to unfold. Buried in thickening sediment, Yorick's jaw became impregnated with minerals seeping into them from the groundwater. It became a fossil.

Seasons turned into centuries, centuries into millennia. Lake Turkana grew again, and shrank. The streams that fed it meandered across the landscape. With majestic slowness, the Rift itself moved. Over millions of years, the sediments that were once the beds of rivers were crushed, stretched, warped and tilted into new positions.

What was once a river bed rose high above the modern water table, and became exposed to the elements. Eventually, the body of sediment containing Yorick's jaw came to rest in a bank of high ground, eroding into the catchment of the Topernawi. Heat shrank the surface, cracking it. Rainwater percolated into the cracks, loosening them. Sun and rain together turned the hard sediment into powder; rain washed and wind blew the dust away. In this way, erosion brought Yorick's jaw to the surface for the first time in more than three million years. Its exposure was a shock to its centuried equilibrium. Unaccustomed to the air and light, the jaw itself began to erode. The heat and desiccation of sunlight strained the surface of the fossil. After millions of years at an even temperature, the outside of the fossil was suddenly markedly hotter and drier than the inside. Layers of bone shifted and exfoliated, peeling back from the teeth, exposing the roots. The bone began to pull itself to pieces.

Inside the jaw, particles of clay that had become lodged in crevices in the gums, beneath the roots of the teeth and in the teeth themselves, heated up and expanded. The pressure they exerted was too much for the fragile, heat-stressed fossil bone to bear. The jaw exploded, shattering into chaff and showering fragments of bone and tooth over several square metres. Later that day, the hooves of a flock of goats crushed most of the identifiable fragments into powder, and dispersed most of the rest still further.

After thirty-three hundred millennia of calm, this destruction happened within the fossil's first few hours after exposure, before the fossil could be seen by human eyes. Presumably, this goes on all the time. With fossil-hunters on site only a few hours in a given year, the irreparable loss of fossils destroyed on exposure, after uncounted intervals of safe burial, is unknowably great.

Yorick's remains wait, bruised by time, for a passing fossil-hunter. But palaeontologists are easily distracted. For each real fossil there are so many false possibilities. Until you pick it up, it is hard to know whether the fossil you think you see is really a pebble of grey Pliocene basalt, or a piece of basalt from the preceding Miocene epoch, as green as jade and studded with purple pyroxenes. Perhaps it is really an eroded wedge of sandstone, gnarled as a ginger root; a slab of bright white carbonate; a powdery cobble of pumice, loaded with feldspars; a shiny calcite crystal, glinting in the light; a lump of friable brown clay; a curve of white chalcedony, or a quartz geode. After three hours of this all you see are pebbles before your eyes. It is then that you start looking for a place to sit down, have a drink, scratch your mosquito bites, and refocus your eyes.

It is at times like this, when you are not really looking, that you find a fossil like Yorick, when you happen to see a glint of blackened enamel out of the corner of your eye, when the angle of light is right, and when the half-molar tooth happens not to be hidden from view by a stone, or the shadow of a thorn bush. Had you arrived yesterday, Yorick might still have been buried with no clue to his presence. You would have taken your break unaware of the prize a centimetre underfoot. Had you arrived the following day, Yorick's half-tooth might have been crushed into unrecognisable particles, shattered by the heat and dispersed by the wind. Through nothing but chance built upon chance, a single tooth has survived a chasm of time beyond human comprehension to tell that its owner once lived.

The fact is that we know so little of the past. We depend on the minute fraction of the life on Earth that has left any record. We have hardly begun to count the species with which we share this planet, yet for every species now living, perhaps a thousand, or a million, or a thousand million – for we will never know for certain – have appeared and become extinct.

In the foreword to *2001: A Space Odyssey*, Arthur C. Clarke and Stanley Kubrick wrote that behind every human being now living stand thirty ghosts. Clarke, recollecting the statement many years later,[16] felt that there ought to have been a good deal more than thirty. The precise number hardly matters. The point is that although these ghosts stand for real ancestors, we have no trace of their graves, their corporeal remains, their diaries, or their headstones.

A few of the Earth's uncounted species evolved into other species. Of these, a very small number indeed established species-on-species genealogies, dynasties so enduring that millions of years later, if we are lucky, we can pick up a fragment here and there to discern the faintest traces of their echoes in the rocks. All but a handful lived and died without leaving us any notice of their existence.

In Shelley's sonnet *Ozymandias*, a traveller in an 'antique land' reports seeing, in the desert, the ruins of a gigantic statue from antiquity. All that is left is a pair of 'vast and trunkless legs of stone', a detached head, half-sunk in the sand, and an inscription. That colossal statue and its inscription are all the evidence that Shelley's traveller had that Ozymandias existed. The inscription proclaims an industrious builder of monuments, but no memorial is left save the decaying statue itself. And what of his conquering armies, his ancestors, his descendants, the thousands of slaves who laboured and died just so the memorial could be built? We have a name and little else. The inscription gives no clue about how long Ozymandias lived, or when, whether he died in the arms of a concubine or beneath the wheels of a usurper's chariot. And yet without any inscription, the story would lose much of its force. Without the inscription, the dismembered statue would be nameless, anonymous, free to float in time, uncomprehended. Had the traveller passed that way a few years later, there might have been nothing to write home about. All trace of Ozymandias, for all his might, would have been wiped from the face of the globe as if he had never been.

The moral of the sonnet is generally taken to concern hubris. How boastful Ozymandias must have been to challenge the gods – and look, how the passage of time has laid his memory low! To a palaeontologist, the sonnet explores the poignant contrast between time as we experience it in our everyday lives and the incomprehensibly greater scale of Deep Time.

The lucky fossil-hunter is a modern version of Shelley's traveller, in a land more than a thousand times as antique. To find one tooth, or even half a tooth – when, by the lottery of life and death, no fragment should have lasted long enough for the finder to contemplate – is surely to be favoured by fate.

The conventional portrait of human evolution – and, indeed, of the history of life – tends to be one of lines of ancestors and descendants. We concentrate on the events leading to modern humanity, ignoring or playing down the evolution of other animals; we prune away all branches in the tree of life except the one leading to ourselves. The result, inevitably, is a tale of progressive improvement, culminating in modern humanity. From our privileged vantage point in the present day, we look back at human ancestry and pick out the features in fossil hominids that we see in ourselves – a bigger brain, an upright stance, the use of tools, and so on. Naturally, we arrange fossil hominids in a series according to their resemblance to the human state. *Homo erectus*, with its human-like upright stance and big brain, will be closer to us than *Ardipithecus ramidus* or *Australopithecus afarensis*, which had smaller brains and more ape-like features.

Because we see evolution in terms of a linear chain of ancestry and descent, we tend to ignore the possibility that some of these ancestors might instead have been side-branches; collateral cousins, rather than direct ancestors. The conventional linear view easily becomes a story in which the features of humanity are acquired in a sequence that can be discerned retrospectively; first an upright stance, then a bigger brain, then the invention of toolmaking and so on, with ourselves as the inevitable consequence.

New fossil discoveries are fitted into this pre-existing story. We call these new discoveries 'missing links', as if the chain of ancestry and descent were a real object for our contemplation, and not what it really is, a completely human invention created after the fact, shaped to accord with human prejudices. In reality, the physical record of human evolution is more modest. Each fossil represents an isolated point, with no knowable connection to any other given fossil, and all float around in an overwhelming sea of gaps.

When Darwin was thinking about mechanisms for evolutionary change, one of his problems was the lack of sufficiently long intervals of time for his schemes to work. He saw evolution as generally slow and gradual, yet the scholarship of his time viewed the Earth as no older than the few thousand years allowed in the Bible. The realization by Victorian geologists such as Darwin's geological mentor, Charles Lyell, that the Earth was inconceivably if not immeasurably old, gave Darwin's idea of natural selection time enough to change one species into another, all the way from the primordial slime to the flora and fauna we see around us today.

In the *Origin of Species*, Darwin put the case for natural selection – his mechanism of evolution – by analogy. Given a group of creatures varied in shape, behaviour and other attributes, natural selection picks those variations best suited to the prevailing environmental conditions, in the same way that pigeon-fanciers select the animals whose features are closest to the desired traits, and use these animals as breeding stock. Give a pigeon-fancier a well-stocked pigeon-loft and enough time, and he could produce pigeons as varied as pouters, tumblers and fantails. By analogy, give Nature a palette of protoplasm on the early Earth and the full span of geological time, and she could produce pigeons, pigeon-fanciers, and everything else.

The analogy between rearing pigeons and natural selection is, however, incomplete. Pigeons bred to be tumblers, pouters or fantails are still pigeons. At no point does the breeder produce a breed of pigeon that is so extreme that you can no longer consider it a pigeon. In Darwin's analogy drawn from fanciers' records, endless varieties can be produced but in no case are new species formed. Artificial selection takes place against the continuity of ordinary, everyday time. Natural selection as devised by Darwin – the force that changes one species into another – does not happen within this timescale.

If this assertion seems rather sweeping, take a look at your own ancestry. Look again at those old family photographs: do any of your ancestors look any more ape-like than you? No, they do not. But just suppose they did – would you expect any ape-like traits to be more marked the further you look back in time? Of course not. If, like me, you are unable to trace your ancestry back more than three or four generations and therefore cannot test these wild and distasteful notions, you might seek solace in more illustrious pedigrees. Queen Elizabeth II, for example, can trace her ancestry back more than a millennium, past the Norman Conquest, back to the Anglo-Saxon chieftains of Wessex. Yet there is no evidence that Alfred the Great was any less human or more ape-like than the present sovereign. So where does this evolution happen?

Let us look at the problem another way. My cat, Fred, is a fine specimen of an oriental lilac siamese. As a product of artificial selection by cat-fanciers, he has the pedigree to prove it. His ancestry can be traced to his great-great-great-great-grandparents, all 32 of them, and those are only the ones whose names can be written comfortably on the breeder's certificate. Some of these ancestral cats were show champions, officially the finest examples of their breed. No doubt, then, that Fred's ancestors are cats – and cats by definition, with certificates and rosettes to prove it. Fred's pedigree, like mine, while perhaps showing signs of artificial selection, shows no sign of evolution, either. Why?

In ordinary, everyday time, the generations of Fred and me run on the separate, parallel tracks, respectively, of feline and human generations, neither converging nor diverging. But in Deep Time, species are malleable, and the parallel lines converge. But how can

33

this be, if my ancestors were all human, and Fred's were all cats?

In ordinary time, organisms such as people, pigeons and cats breed true to their kind. Deep Time, in contrast, is time enough for species themselves to transmute. Deep Time is the key to the origin of species, for intervals of time of geological extent were required for Darwin's mechanism – natural selection – to do its work and change one species into another. Evolution is a consequence of Deep Time.

Although Fred and I are different in many ways, we have much in common, signs of a shared evolutionary heritage. At some unspecified location in Deep Time, there was a creature, our latest common ancestor, from which the lineages leading to Fred and me diverged, each going on its own, separate course.

The problem is that Fred and I cannot place our common ancestor in time and space unless we are able to discover our complete pedigrees all the way back to that point of ancestral convergence. To do this, as we know, is impossible, given that the fossil record is so discontinuous. All we know is that such an ancestor existed – sometime, we know not when; somewhere, we know not where. It is conceivable that we could dig up her fossil remains but even if we did, we could never know that we had done so. Even so, we can still get some idea of what our latest common ancestor was like, even without fossil evidence.

The evidence of evolution is everywhere around us, in the signs that diverse organisms share a common morphological heritage. That Fred and I have a common ancestry is not in dispute, not because of fossils, but because of features we share thanks to our common evolutionary birthright.

For all our superficial differences, Fred and I are very similar underneath. We both have backbones made of stacks of vertebrae, and skulls enclosing our brains and protecting our paired eyes and ears. We share these features thanks to a common ancestry going back more than 500 million years, to when the first vertebrates – backboned animals – appeared in the sea. The earliest fossils of vertebrates are just scaly flakes and fragments; even so, we know, just by looking at Fred and myself, and knowing that all vertebrates have a backbone, a skull and paired sense organs, that our common ancestor must also have had these features.

Yet Fred and I share features indicating a more exclusive common ancestry than that indicated by the shared possession of the characteristics of vertebrates. We both have tooth-bearing jaws, and two pairs of fleshy limbs. The shared heritage of limbs in this pattern marks Fred and myself as not only vertebrates but 'tetrapods'. All tetrapods are vertebrates, but not vice versa; we have two pairs of limbs in addition to the backbone, skull and sense organs, making us members of a more exclusive group. The latest common ancestor of Fred and myself must have had these shared features, acquired along evolution's journey.

But Fred and I share yet further characteristics, indicating a still more exclusive common ancestry. Not only do we have a backbone, a skull, paired sense organs, four limbs, and so on; we also (thanks to our shared heritage) benefited, as embryos, from the protection of membranes that saved us from drying out, kept us fed, and looked after our bodily wastes. One of these membranes is called the 'amnion' so Fred and I are both 'amniotes'.

Our common heritage is more exclusive still: we both have hair, and a highly regulated metabolism that keeps a constant body temperature. We also share a lot of little details in our anatomies, such as the precise way that our jaws hinge onto our skulls. We also, when we were very young, suckled milk from our mothers from organs provided for the purpose – mothers who had previously nurtured us in wombs, rather than laying eggs (which some amniotes, such as reptiles and birds, generally do). By virtue of all these shared features, both Fred and I are mammals.

Our latest common ancestor must also, therefore, have been a mammal. Long ago, in some long-lost crevice, a mammal had a litter of progeny, which went their separate ways to seek their fortunes. One founded a lineage whose end result is me. Another multiplied to produce the dynasty of which Fred is the latest scion.

When did our most recent common ancestor live? Where? What was she like? We cannot reliably know the answers to the first two questions, but we can get some idea of the third, simply by comparing the features that Fred and I share. She was a warm-blooded furry animal that bore live young and suckled them.

We cannot paint a more complete portrait. Although our common ancestor had four legs and a tail, we cannot know how long her tail was, or the colour of her fur. Although we know that she had paired eyes and ears, we don't know whether those eyes were blue, green, brown or pink, or if her ears were rounded or pointed. Although she had teeth, we do not know if they were like Fred's feline fangs, my own more modest molars, or like something else instead, of a shape that no longer exists in any living animal. Perhaps our common ancestor was a rat-like creature. Or perhaps not. We will never know, because, in the intervals of Deep Time since those long-lost litter-mates left the nest, evolution gave each lineage its own distinct inventory of features, the history of which explains why Fred is a cat, and I am a human. This can all be summarised in a simple diagram (Figure 2).

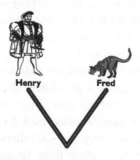

Figure 2. How the ancestries of Fred and myself are linked, through Deep Time, with our common ancestor at the 'node'.

This is not a genealogy; it is not an explicit family tree. All it does is summarise the tale about the acquisition of features, showing that Fred and I represent two separate lineages that diverged beyond a 'node'. There is an important distinction to be made between this diagram and an explicit genealogy. In a genealogy, the node would represent a real individual.

In this diagram, the real individuals – the ones we can know about – appear at the tips. The node represents an ideal state; not the common ancestor by name but the inventory of shared features, acquired in evolution, which we would expect it to have, given what we know about Fred and myself.

That Fred and I really did have a common ancestor is not in doubt, but we cannot hope to find her as a fossil; or if we were to find her, we could never know for certain that we had done so. This principle is reflected in the diagram, in the form of the node linking my ancestry with Fred's. The node reflects a notional embodiment of the inventory of features shared by Fred and myself – and that is all. It says nothing about when or where our latest common ancestor actually lived.

Neither can the diagram tell us about any features that made that common ancestor distinctive in her own right. For all we know, our shared common ancestor had features peculiar to herself, none of which are present in either Fred or myself, and which are lost for ever. All the node does is mark a waypoint in Deep Time, the point when the heritage shared by Fred and myself gave way to separate and divergent histories. Importantly, it doesn't matter whether the lines linking me, Fred and the node are short or long, or of different lengths or thicknesses. It doesn't matter whether the angle joining them is obtuse or acute. All that matters is the way they are joined; the topology.

Having established all that, the diagram in Figure 2 doesn't do anything besides express a truism, that Fred and I have a common ancestor, and that we are cousins in some unknown degree. If life had a single origin, then every organism that ever existed, alive or dead, is a relative of every other organism, alive or dead. Even if we can never know that Yorick, the fossil I hold in my hand, is my direct ancestor, we can be *sure* that he is my cousin, yours, and Fred's. You can, with equal facility and justification, find any two creatures, living or extinct, and draw a diagram as in Figure 2 to express their shared common ancestry. In which case, a diagram expressing the common ancestry of any two living creatures is always true.

There is, however, a more constructive way to explore the relationship that I have with my cat Fred. Rather than state that any two organisms are cousins to an unknown degree, we can try to estimate this degree, at least in relative terms. To do that, we need to introduce some perspective in the form of another participant.

My other cat, Marmite, unlike Fred, has the breeding and manners of an alley cat. Her pedigree is even less well-established than mine, let alone Fred's. Even so, Marmite and Fred have a lot in common. Like Fred, she has pointed ears, whiskers and retractile claws, likes to chase mice and to tease the dog next door. I can claim none of these attributes.

However, all three of us like to eat fish and go to sleep in the afternoon when we have the chance. More fundamentally, we are all vertebrates, tetrapods, amniotes and mammals. But because Marmite and Fred have more in common with each other than either does with me, we can reasonably infer that Marmite shares a common ancestry with Fred from which I am excluded; as cousins, she and Fred are closer to each other than either is to me. To put it another way, the latest common ancestor of Fred and Marmite lived later than the latest common ancestor of all three of us, although we cannot know how much later. Figure 3 expresses our mutual relationships.

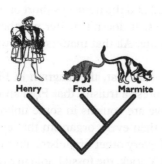

Figure 3. A diagram showing the inferred relationships between myself and my cats, Fred and Marmite.

The diagram in Figure 3 says more than the truism expressed in Figure 2. This is because it makes a particular statement about the order in which organisms are related, over and above the fact that they all share some degree of common heritage. In a topological sense, the diagram in Figure 2 must always be true because there is only one way to draw it. In contrast, there is more than one way of arranging three (or more) participants in a diagram like this, so the statement about relationships in Figure 3 can be tested against possible alternatives.[17]

This notion of testability is crucially important, because it elevates the diagram from an assertion to the status of a scientific hypothesis. A diagram of this sort is called a 'cladogram', from a Greek word meaning a branch. The business of drawing up and testing various alternative cladograms is called 'cladistics'.

In cladistics, presumptions about particular courses of ancestry and descent are abandoned as unprovable or unknowable. Yet cladistics does more than state that we are all cousins. It is a formal way of investigating the order in which organisms are cousins, by examining the possible alternatives. Cladograms are statements of collateral relationship of greater or lesser extent. Given that, they sidestep the question of whether Yorick is my ancestor, or whether any fossil is the ancestor of any other; for the answer to these questions can never be known. In other words, cladistics acknowledges the discontinuities of Deep Time and, by acknowledging them, transcends them.

Figure 4 illustrates one of the alternatives to Figure 3 – an alternative cladogram – for the relationship between myself, Fred and Marmite. In this cladogram, I share a common ancestry with Fred that excludes Marmite.

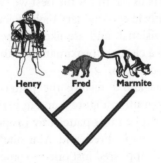

Henry   Fred   Marmite

Figure 4. An alternative relationship between myself, Fred and Marmite.

If it is possible to arrange three participants in more than one way, how can one know which one reflects the actual course of evolution – what really happened? The cladogram in Figure 3 seems so natural, so plausible, that it must be right. But must it? The fact that three participants can be arranged in a different way from that intuitively expected suggests that it is at least possible to conceive of different evolutionary courses. To dismiss this, and assert that the cladogram in Figure 3 must be right simply because it accords with native common sense, is not a scientific approach, because it allows only what we humans imagine is possible, and denies us the opportunity of exploring all the alternatives and examining which one best fits the evidence at hand.

How, then, is it possible to evaluate the likelihood of different alternatives? Formally, it is not, because we cannot discover the particularities of ancestry and descent that join us together. In practice, however, we can adopt a principle that has stood science well for centuries. That is the 'Principle of Parsimony', or 'Occam's Razor'; when two hypotheses present themselves, it is prudent to choose, as a working hypothesis, the one that requires the fewest assumptions to justify itself. It is important to realise that the principle of parsimony does not select the 'right' answer – for that is unknowable – but only the best one to be getting on with first. Because we cannot hope to retrieve the continuous skein of ancestry and descent that links us all, this is the best we can hope to achieve.

This is true not only in evolutionary biology, but in all science. All hypotheses are provisional, and are likely to be overthrown when new evidence allows a closer approximation to the truth. If this were not true, science would stop.

How can my cats and I resolve the pattern of our heritage? Looking at the cladogram in Figure 3 and applying the principle of parsimony, we can explain this pattern by proposing that all those feline features that define Fred and Marmite's ancestry, to the exclusion of my own, appeared just once – somewhere between the node bracketing the two cats, and the node that defines all three of us.

The situation in the cladogram in Figure 4 is more complicated. If Fred and I share a common ancestry that excludes Marmite, then the cat-like features that Marmite and Fred so evidently share would have evolved twice, independently, in the separate lineages leading respectively to Marmite and to Fred.

Because the cladogram in Figure 3 implies that cat-like features evolved only once, and the one in Figure 4 implies that they might have evolved twice, parsimony favours the first. As a working hypothesis, it is simpler to imagine that all the features so characteristic of cats – the pointed ears, the whiskers, the retractile claws, the constant demands to be let into the garden (and readmitted thirty seconds later) and so on – evolved just once, rather than twice independently. It is entirely possible that these features really *did* evolve independently, on more than one occasion – nobody says that evolution is parsimonious – but, for the sake of simplicity, we choose the most parsimonious alternative as the one to be getting on with, for the time being, until some other evidence turns up to favour a different view.

The problem is actually more complicated, because the cladogram in Figure 4 has a second interpretation. That is, the common ancestor of myself, Fred and Marmite was cat-like, and these features have been lost, just once, in the lineage leading to me. This second interpretation is just as parsimonious as the cladogram in Figure 3, except that features are lost, rather than acquired. How can we decide between the two?

Sometimes it is not possible to decide between two equally parsimonious cladograms. In this case, we can break the deadlock by introducing a further element of perspective, by testing the two alternatives against a fourth participant which, based on independent evidence, we have good reason to suspect stands outside the three of us. In cladistics, this kind of arbitrator is called an 'outgroup', and the process of arbitration, 'outgroup comparison'.

I elect to use, as an outgroup, one of the pigeons that Marmite often chases out of her London garden. This is not an arbitrary choice. Like myself, Marmite and Fred, the pigeon is a vertebrate, a tetrapod and an amniote. However, unlike people or cats, the pigeon lacks mammalian features such as hair, mammary glands and the habit of suckling infants. From this, we can assume that the latest common ancestor of the pigeon and the rest of us did not have mammalian features. Although the pigeon is a cousin to Marmite, Fred and myself, we have good reason to think that it stands at a remote remove.

Figure 5 shows the cladograms in Figure 3 (on the left) and Figure 4 (on the right) with the pigeon added as an outgroup. The cladogram on the left requires the evolution of cat-like features only once, as it did in Figure 3, in the branch leading exclusively to Marmite and Fred. However, the cladogram on the right requires *either* that these cat-like features evolved in the common ancestry of Marmite, Fred and myself, but were subsequently lost in my own ancestry, *or* that they were acquired independently in the separate lineages leading to Marmite and Fred. Either there is a gain of cat-like features and a subsequent loss of all these features, making two events – a gain and a loss – or cat-like features were gained twice, independently, again making two events. Because they each contain two events, both interpretations of the cladogram on the right are less parsimonious than the one offered in the cladogram on the left, which requires just one event. Once again, the cladogram in Figure 3 comes out as the best provisional hypothesis. As far as we can tell on the evidence we have, Fred and Marmite share a common ancestry that excludes me.

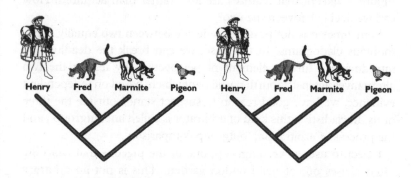

Figure 5. Adding the perspective of an outgroup (a pigeon, in this case) allows the resolution of two equally parsimonious cladograms.

I am now going to do something rather mischievous; something that only cladistics allows. In Figure 6 I add a fossil to the panoply of living, breathing entities in the cladogram I drew up in Figure 3.

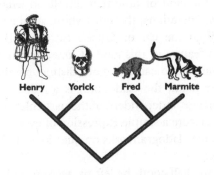

Figure 6. A cladogram with a fossil in it.

This cladogram is just like the one in Figure 3 inasmuch as it concerns me and my cats. It is different in that it adds Yorick – the imaginary hominid fossil – and places him next to me. The implication is that Yorick and I share a common ancestor quite different from the latest common ancestor of Marmite and Fred. The common ancestor of Yorick and myself would have had features that one does not see in cats, such as the peculiarities of hominid dentition. Even though Yorick is now less an individual than a chipped tooth, we know enough to see that the tooth is more like my molars than Marmite's or Fred's fangs.

Something about this cladogram is astonishing. It is simply that three living individuals – Marmite, Fred and myself – are found in the cladogram, on the same plane, on the same footing, as a fossil, Yorick, who is not only dead, but has been so for more than three million years. This cladogram transcends time.

In a conventional genealogy, relationship is plotted from side to side, across the page, whereas time is drawn from the top of the page to the bottom. Conventional genealogies, then, are like the embossed rolls in a player piano; the notes are arranged in time according to the specific places where they appear on the roll as it unwinds. This cladogram seems to ignore time entirely; all it does is plot degrees of evolutionary relationship.

But it is more subtle than that. My cladogram ignores only ordinary time, the kind of time that unrolls in lengths of days or years, or a few generations, the time whose span encompasses the ancestries of Fred the cat or Queen Elizabeth II. Evolutionary change is not a feature of this kind of time, in which we descend, human to human, cat to cat, along our distinct parallel tracks.

On the other hand, the cladogram is a suitable expression of evolutionary change, independent of the particulars of ancestry and descent. It is therefore a simple expression of evolutionary relationships. As such, the cladogram asks no questions of Deep Time that it cannot answer.

Yorick, and the half-tooth he left us, were products of imagination. Yet the fragment of jaw that Robert found beneath a thorn tree was real enough, as was the sheared molar that Gabriel unhesitatingly drew from the sieve before me. These are real fossils that were once part of real individuals that existed in time and space. We are entitled to say that the fossils discovered at LO5 were our cousins to some degree – yours, mine, Gabriel's, Robert's, Nzube's, Meave's – but not to assert, without question, that they were, or might have been, our direct ancestors.

No matter how fragmentary or antique, the bones and teeth that Meave Leakey's team continues to unearth belonged to real individuals that walked the Earth as surely as Fred, Marmite and I do today. If we cannot imagine ourselves as missing links, we are not entitled to demand this of the fossils we find. They were neither missing links nor transitional forms: they existed in their own right, not as staging posts in a preordained story between the apes and the angels.

2 Hunting Unicorns

> In such conditions, we would be face to face with a unicorn and not know for certain what it was. We know that such and such an animal with a mane is a horse and that such and such an animal with horns is a bull. But we do not know what the unicorn is like.
>
> Jorge Luis Borges, *Kafka and His Precursors*

Fossils are mute. Unable to tell us their own stories, we tell their stories for them, to flatter our own prejudices: tales of noble lineage, of brave quests against adversity, of battles against formidable enemies, and successions of triumph on triumph, culminating in mankind – ourselves.

Figure 7. 'I will be the *First* on Land': a frame from Larry Gonick's *Cartoon History of the Universe*.

One of these tales concerns the conquest of the land by our aquatic ancestors, some time before 360 million years ago. '*I* will be the *First* on Land', declares an ancestral fish as it heaves itself ashore, in Larry Gonick's *Cartoon History of the Universe*.[18] 'Hmmm ... appears the *bugs* are already there', observes a cautious colleague coming up behind. 'They won't get the credit', booms the first – '*my* descendants will write the *book*'.

This is a gentle send-up of all the comforting tales we like to tell ourselves about our past, about our ancestors, the fishes, boldly climbing onto land. Those fishes had all the qualities of industry, invention and initiative that we, as human beings, admire in ourselves: after all, they were *our* ancestors. And as Gonick also wrily observes, the victors get to tell the story. Had the bugs turned out to have been historians, the tale of conquest might have been very different. Gonick's satire succeeds because the traditional tale about how those fishy pioneers made their heroic journey from their watery homes, boldly venturing into the new and hostile terrestrial environment, is ingrained in popular culture. The cartoon is a play on familiar human tales of conquest similarly subject to later revisionism, such as the opening up of the American West. Gonick's fishes are cowboys or conquistadors; the bugs are the indians, the aborigines, the displaced indigenes.

I first discovered our fishy ancestors when I was about five years old, and my parents took me to the Natural History Museum in London. In my heart, I have never left its grand, Victorian halls. In every gallery, cases spanned the floor and climbed the walls, old wood and glass, crammed with incomprehensible specimens and all-but-unreadable labels. Their inaccessibility only added to the atmosphere.

My favourite gallery was one of the least visited – the Hall of Fossil Fishes. More cases of glass and wood, and labels bearing incredible words – antiarch, teleost, arthrodire, *Petromyzon*, amphioxus, heterostracan, Osteichthyes, rhipidistian, Crossopterygii, ostracoderm, *Jamoytius* – words as potent as any incantation. Along one side of the gallery was a series of what looked like fish tanks set into the wall, full of painted plaster models of the fossil fishes in life, in their long-vanished habitats.

Among these fishes, somewhere, were the ancestors of the tetrapods – the land vertebrates that now include you, me, Fred the cat, and most familiar creatures. All tetrapods – that is, all amphibians, reptiles, birds and mammals – descend from that first finny adventurer above the slime.

Today, there are more species of fishes than of amphibians, reptiles, mammals and birds put together. In the great scheme of things, tetrapods represent a small group of specialized fishes. Some fishes are specialized for living in the ocean depths, under a column of water miles thick: so much so that when they are brought closer to the surface, they explode from decompression. Other fishes are specialised for life in water so shallow that the water sometimes disappears altogether. These are the fishes we call 'tetrapods'. Because we are also tetrapods, we naturally attach a disproportionate importance to the origins of this particular subgroup of fishes. But which, among the panoply of other fishes, are the ancestors of the tetrapods?

I discovered an answer in the Hall of Fossil Fishes. The displays told me that the tetrapods sprang from a greater sub-group of bony fishes called the Sarcopterygii, the so-called 'lobe-finned' fishes, whose paired fins are held clear on lobes of flesh supported by bones. This ancient group was once diverse, but as in many aristocratic families, the passing eons have brought eccentricity, etiolation and extinction. Sarcopterygians were once common, but of the thousands of different kinds of fishes in the modern world, only four remain (that is, if one excludes the tetrapods, which are sarcopterygians specialised for land life). Evolution has wrought great changes in the forms of these fishes over the course of almost 400 million years. In their current state, specialised for peculiar modes of life, these fishes need tell us very little about their evolutionary relationships, or what their ancestors looked like.

Three of the four surviving non-tetrapod sarcopterygians are the so-called 'lungfishes', one each from Africa, Australia and South America. The African and South American forms are elongate and snake-like, with wispy, string-like ciphers of fins. Only the Australian form has fins fleshy enough to merit its designation as a lobe-finned fish. They live in fresh water but, when the water dries out, the African and South American forms can cocoon themselves in burrows until the rains come again, in the manner of some desert amphibians.[19] Indeed, when lungfishes were first discovered, they were thought to be amphibians, not fishes. But, as it turns out, the amphibian-like habits of modern lungfishes reflect specialisations acquired relatively recently, not any close relationship with tetrapods. Fossil lungfishes look a great deal more fish-like than any modern form, and are known to have lived in the sea as well as fresh water. The extant lungfishes remain our distant cousins, not our ancestors.

The fourth extant non-tetrapod sarcopterygian, like the Australian lungfish, looks like a genuine throwback to a bygone era. Since the mid-nineteenth century, palaeontologists have been familiar with a group of fossil lobe-finned fishes called coelacanths. These appeared in the Devonian Period (408 to 360 million years ago), had their heyday in the Triassic around 220 million years ago, and slowly declined thereafter. The most recent coelacanth known exclusively from fossils died out around 70 million years ago, when dinosaurs still ruled the Earth.[20] As a group, coelacanths were very 'conservative'. That is, they all looked very similar, the earliest species looking much like the last ones known from the fossil record, living 300 million years later. When palaeontologists first cast their notional nets over the side of Deep Time, coelacanths were as extinct as a host of other lobe-finned fishes.

There seemed no reason to think otherwise until the carcass of a recently dead coelacanth was recovered off South Africa in 1938, after a gap of 70 million years.[21] The discovery caused a sensation: just imagine the consternation that might attend the discovery of a recently dead road-kill of the dinosaur *Velociraptor* on a dirt track in Mongolia. Yet *Velociraptor* was a near-contemporary of the last fossil coelacanth.

There was a frustrating wait for the second coelacanth, which turned up in the Comoro Islands in the Indian Ocean in 1952. Since then, more than 200 have been caught in the Comoros, though some have been spotted off nearby Madagascar. The discovery of another population 10,000 kilometres away, off Sulawesi in Indonesia in 1998, was the zoological surprise of the year.[22]

Unlike the modern lungfishes, the modern coelacanth looks very similar to its fossil relatives. Again, in contrast to the lungfishes, which are freshwater fishes, the coelacanth is an inhabitant of the ocean depths. This 1.5-metre, 45-kilogramme fish prefers to live in and around caves in sheer submarine cliffs at a depth of around 200 metres. Like all sarcopterygians, lungfishes included, the coelacanth has a lung – but it is full of fat and quite useless for breathing air.

The lungfishes and coelacanth are small vestiges of a formerly much greater diversity of lobe-finned fishes, ranging from huge and fearsome-looking lobe-finned fishes called rhizodonts, to the peculiar onychodonts, still largely untouched by research. All these wonderful creatures have been extinct for hundreds of millions of years.

Although these lobe-finned fishes were diverse in size, habit and habitat, they all shared a number of distinctive features. They had big, heavy scales and fleshy fins supported by rods of bone that correspond to the long bones in tetrapod limbs. The pectoral fins of most fishes (the front pair, immediately behind the head) stick onto a bony plate in the body wall. In lobe-finned fishes, each pectoral fin is held clear on a bone called the humerus that branches, further down the limb, into a radius and an ulna. These bones correspond with the humerus, radius and ulna in the arms of tetrapods, including you, myself and Fred the cat. The pelvic fins, the pair of fins at the back, are likewise each supported by a femur, a 'thighbone', that branches into a tibia and a fibula, before branching further into a fan of small 'radials' that support the fin membrane itself.

Among all the lobe-finned fishes, the ancestry of the tetrapods was traditionally pinned most closely on yet another of the extinct groups of lobe-finned fishes, the osteolepiforms. One osteolepiform fish in particular figured prominently in discussions about tetrapod origins. This was a creature called *Eusthenopteron*, an impressive fish that occasionally reached a metre in length, whose body finished rather grandly in a three-pointed tail, like Neptune's trident. Apart from its fleshy, bone-supported fins, the skull and teeth of *Eusthenopteron* resemble those of early fossil tetrapods in some detail. Between 1942 and 1954, a Swedish researcher named Erik Jarvik published a series of papers on the anatomy of this animal. As a result, *Eusthenopteron* became by far the best-known of all osteo-lepiform fishes, its anatomy being better understood than that of many living, breathing creatures.

Because osteolepiforms were seen among all the fishes as the group most closely related to tetrapods, and because *Eusthenopteron* was the best characterised osteolepiform, *Eusthenopteron* made its way into popular accounts as the closest thing there was to the ancestor of tetrapods. From there, the elision was easy, and whenever discussion turned to tetrapod ancestry, *Eusthenopteron* was paraded into the picture. Less often stated was the fact that *Eusthenopteron* was the fittest fish only because nothing more suitable was available, or had been studied in such detail. Because it was the best-described osteolepiform, it was assumed to have been a typical representative of the group from which tetrapods originated and thus a suitable candidate for a tetrapod ancestor. Had Jarvik chanced to study some other osteolepiform instead, that fish, and not *Eusthenopteron*, might have taken on the mantle of tetrapod ancestor.

Moreover, nobody had ever defined what was meant, precisely, by an 'osteolepiform', in terms of an explicit list of features that osteolepiforms shared exclusively among themselves. The osteo-lepiforms were less a natural group than a mongrel assortment of somewhat unspecialised lobe-finned fishes, with no particular features to make them out as different from any of the others. To say, therefore, that the tetrapod ancestor was an osteolepiform is actually meaningless. If it is not clear what an osteolepiform is, you cannot distinguish one from any other lobe-finned fish. If any of these reservations were written on specimen labels on my early visits to the Natural History Museum, back in the 1960s, their importance was lost on my five-year-old mind. What I took away was a child's mental picture of *Eusthenopteron* as a big green fish (colour, of course, conjectural) with a benign and knowing smile, on which was pinned a bright red rosette labelled 'Our Ancestor'.

What were the first tetrapods like? In several expeditions to east Greenland in the 1930s, a team of Swedish and Danish explorers unearthed the bones of what were then the earliest and most primitive known tetrapods, from rocks now known to have come from the very end of the Devonian Period, about 360 million years ago. The task of describing these finds in the scientific literature fell to a talented young palaeontologist, Gunnar Säve-Söderbergh, who wrote a preliminary description of the most abundant tetrapod, which he named *Ichthyostega*. The archaic appearance of the animal was not lost on the public of the time; when only parts of the skull had been described, a Danish cartoonist celebrated *Ichthyostega* as a herring with legs, and dubbed it *den firbenede fisk* – 'the four-legged fish'.

But Säve-Söderbergh's efforts to write a full, monographic description of the finds were cut short. He died in 1948 of tuberculosis, aged 38, the work on *Ichthyostega* incomplete. The job was passed on to a promising student who had been on the Greenland expeditions – the same Erik Jarvik who later went on to study *Eusthenopteron*. Because tetrapods from a time as remote as the Devonian period were so new to science, and *Ichthyostega* looked so different from any other tetrapod, living or fossil, Jarvik felt unable to place *Ichthyostega* as an animal until he had comprehensively redescribed the anatomy and relationships of Devonian lobe-finned fishes. Without this, it would be impossible to set these earliest of tetrapods into their evolutionary context. Establishing this context became Jarvik's life's work. This study led to a detailed understanding of *Eusthenopteron*, and much else of value, but delayed a full description of the tetrapod fossils for six decades. Jarvik's monograph on *Ichthyostega* finally appeared in 1996, a short time before his own death.[23]

Ichthyostega is a curious mélange of fishy and tetrapod features. It has four robust limbs, and a barrel-like chest composed of great, flat ribs that overlap like the timbers in a clinker-laid ship, swaddling the chest cavity in an armoured cage. The body finishes with a flourish, a long tail that supports a fish-like tail fin, complete with fin rays. The name *Ichthyostega* means 'fish roof', referring to the pattern of bones on the top of the skull, which looks broadly similar to that seen in Devonian lobe-finned fishes such as *Eusthenopteron*.

When *Ichthyostega* was discovered, it was the earliest known tetrapod by a large margin. The next earliest came from the middle of the succeeding Carboniferous Period, tens of millions of years later. Although these Carboniferous tetrapods were themselves often rather strange creatures, they lacked the fishy features of *Ichthyostega*, making *Ichthyostega* look truly antique by comparison. *Ichthyostega* thus fell naturally into the role of a missing link between non-tetrapods – fishes such as *Eusthenopteron* – and tetrapods such as amphibians and reptiles and, ultimately, us.

Ichthyostega's status as a missing link persisted for many decades. Even in the 1960s, when I first visited the Hall of Fossil Fishes, *Ichthyostega* remained the earliest known tetrapod, conqueror of the land, with *Eusthenopteron* cast as a kind of prophet, heralding the new dawn of the tetrapods. Nobody asked whether the robes of ancestry that we had asked these creatures to wear fitted them well or ill. Yet *Ichthyostega* and *Eusthenopteron* were just two creatures, two isolated points, that we had chanced to hook from the well of Deep Time. There was no guarantee, or any way of knowing, that these creatures in particular – rather than other animals, unfossilised, undiscovered and unsung – were our ancestors.

To propose a link of ancestry and descent between *Eusthenopteron* and *Ichthyostega*, and, through *Ichthyostega*, to more 'advanced' tetrapods, is to impose our own ideas of the past, retrospectively, on the evidence that chance throws at our feet. This imposition promises a greater danger – of reading the fossil not as it actually *is*, but in terms of what we think it *ought* to be, according to the prescription of our story of ancestry and descent. For example, if we come to *Ichthyostega*, our minds loaded with reverence for the weight that this fossil carries as the earliest tetrapod, we will see *Ichthyostega* only in terms of later tetrapods. We will play down any peculiarities that the creature might have had, peculiarities that bear no relevance either to the character of later tetrapods or to our preconceived story.

A critical look at *Ichthyostega* is something like the study of the unicorn by Han Yu, a ninth-century Chinese writer, as dubiously reported by Jorge Luis Borges in his essay *Kafka and his Precursors*.[24] The unicorn is a creature beyond everyday experience, says Borges; the animal 'does not figure among the domestic beasts'. Were we to confront one, we would not be able to make any sense of it except as a chimera of bits and pieces of more familiar animals. We would see in it something of the horse or the bull, but we would get no sense of the unicorn as a unique whole, on its own terms, as something different. Because we have never seen a unicorn before, we are compelled to see it through an overlay of the familiar.

But just because the unicorn looks something like a bull or a horse to us, this does not mean that a unicorn is a 'missing link' between these two animals. Horses and bulls are contingent – they just happened to offer themselves as models because they are familiar and available. Perhaps, in another part of the world, a unicorn would be seen as a mixture of a camel and a kudu – but a unicorn would not be a missing link between those animals, either. The fact is that we are constrained to interpret unknown forms in the light of the known – and the known is limited by the animals we have around us today, a shadow of past diversity, and by human experience. This thinking, in part, explains why Jarvik embarked on a thorough, decades-long anatomical tour of the fossil fishes, broadening his experience of all the characteristics of the fishes, before attempting the unicorn strangeness of *Ichthyostega*.

A statement that *Ichthyostega* is a 'missing link' between fishes and later tetrapods reveals far more about our prejudices than about the creature we are supposed to be studying. It shows how much we are imposing a restrictive view of reality based on our own limited experience, when reality may be larger, stranger and more *different* than we can imagine.

Here is an example of what I mean. To me, 'fish fingers', those nursery-food staples, have always carried an aura of contradiction: fishes don't have fingers. Fingers, and toes, too, are features unique to tetrapods. But how many digits did the first tetrapod carry?

Generations of zoology students have been taught that the first tetrapod must have carried five digits on the end of each limb. The reasoning is simple: most tetrapod limbs, including our own, carry five digits. With rare exceptions that can be put down to congenital malformation or subsequent injury, we all have five fingers on each hand, and five toes on each foot. It is possible that digits may be lost during evolution; birds have the vestiges of three fingers, cattle just two, horses only one; but it is easy to work out a pattern of loss starting with the 'primitive' number, five. No normal tetrapod ever has more than five digits.

If *Ichthyostega* were truly a missing link and an ancestral tetrapod, we might assume that it would have had five digits on each foot. But how secure is this assumption? The fossils originally available to Säve-Söderbergh said nothing on this point; another fact played down in the popular tale of terrestrial conquest was that too little was known about the hands and feet of *Ichthyostega* to be absolutely certain how many digits they carried. The hands and feet of *Ichthyostega* and, as it turned out, other Devonian tetrapods, bore surprises kept secret for decades. In the 1980s, a Russian researcher called Oleg Lebedev described a strange Devonian tetrapod called *Tulerpeton*, which had six digits, but nobody took much notice. It took a return trip to Greenland to expose just how strange Devonian tetrapods really were.

Between 1968 and 1970, geologists from the University of Cambridge were working in the eastern part of Greenland that had produced the fossils of *Ichthyostega* in the 1930s. The Cambridge team was principally interested in stratigraphy – the geological mapping of the rocks – rather than the fossils they contained. However, the geologists found tantalising specimens of Devonian tetrapods, hinting that further finds could be made by palaeontologists who knew what they were looking for. Most of the intriguing hints concerned a little-known contemporary of *Ichthyostega*, a tetrapod called *Acanthostega*. Like *Ichthyostega*, a specimen of this creature, a crushed skull, was first described by Säve-Söderbergh who, perhaps intuitively, realised its strangeness from the start. As Jarvik recalls in his 1996 monograph, 'I still remember Säve-Söderbergh sitting for hours in our tent, twisting and turning this specimen with a puzzled face'.

Jenny Clack, a palaeontologist from the University Museum of Zoology in Cambridge and a specialist in the fossils of early tetrapods, wanted to find more specimens of *Acanthostega* and possibly of *Ichthyostega*. The material from the 1930s expeditions was being minutely studied by Jarvik, and he was still a decade away from publishing his work on *Ichthyostega*. Palaeontological protocol dictates that while one researcher is writing primary descriptions of new material, other researchers cannot study it so intensively that the person doing all the hard work will be pipped to publication.

Clearly, if anything more was to be learned about early tetrapods without waiting for Jarvik to publish, you had to go to Greenland and look for your own material. So, in the summer of 1987, Jenny joined forces with a Danish expedition to eastern Greenland.

The 1987 expedition returned with three-quarters of a ton of rock, filled with fossils of *Acanthostega*; not just fragments, but complete skulls and skeletons, including limbs. Jenny set to work on the fossils, accompanied in her study by a young postdoctoral researcher, Mike Coates. Jenny and Mike published a series of papers on *Acanthostega* through the late 1980s and 1990s, painting a progressively more complete portrait of *Acanthostega*, a tetrapod even stranger than *Ichthyostega*.[25]

One of the first curiosities was the hyomandibula, the equivalent of the tiny stapes ('stirrup') bone in our middle ear. In fishes – and, as it turns out, in *Acanthostega* – the hyomandibula forms a substantial structural element that holds the back of the skull together. At first, Jenny thought that it might have been a kind of paddle, pumping water in and out of a rudimentary gill slit. She later revised that view, adopting the opinion that the hyomandibula was a static structural element, bracing the upper jaw against the braincase, as seen in many fishes, but not in tetrapods. But the fish-like character of *Acanthostega* didn't end there.

The throat region of modern tetrapods such as you and me, and Fred the cat, contains a bone called the hyoid. This anchors muscles that work the tongue. The hyoid is a vestige of an elaborate basket-like structure that supports the internal gills of fishes. Next time you buy a whole, freshly caught salmon or trout, place the fish sideways on a chopping board and pull back the stiff, silvery semicircular flap – the operculum – on the side of the head. You will see, under the operculum, the pink, fleshy gills supported on a series of bones. Remarkably, *Acanthostega* has a set of gill-supporting bones, very similar to what you'd find in a fish bought at your local fish market. Nothing like it has been seen in any other tetrapod, living or fossil. The tadpoles of some amphibians (all of which are tetrapods, of course) retain gills, for breathing underwater. A few amphibians, such as the axolotl, retain these gills as adults. But, importantly, these gills are feathery, external structures, not the internal gills typical of fishes – and, as it turned out, *Acanthostega*, uniquely among tetrapods. The presence of internal gills means that *Acanthostega* was fully and obligately aquatic, completely unlike any tetrapod but perfectly like a fish.

Looking at the rest of the skeleton, Jenny and Mike found that *Acanthostega* had a big, fishy tail, just like *Ichthyostega*. The body of *Acanthostega* was not quite as robust as in *Ichthyostega*, lacking the big, blade-like ribs and barrel chest. But the strangest things of all were the limbs. Each limb had eight digits.

When Mike reconstructed the *Acanthostega* skeleton, piecing together the elements into something like a natural posture, he found that the front limbs could not possibly have worked to support the body clear of the ground. Indeed, they stuck straight out at the sides, and would have been capable of only a limited range of movement, more or less restricted to 'flapping' a little way up and down. The hindlimbs of *Acanthostega* were capable of greater movement than the forelimbs, and could have worked like paddles.

By the time Mike and Jenny had pieced together this remarkable creature, what lay before them was a tetrapod so primitive that it could be considered equally well as an unusual lobe-finned fish – a fish with legs and digits instead of fins. This fully aquatic animal would have been incapable of crawling very far on land. If it did and got stranded, it would have been able, presumably, to use a lung (a feature of many primitive fishes, not just lungfishes and coelacanths) but its internal gills would have dried out. It would have been far more adept in the water, however, paddling with its hindlimbs and tail, using its near-static forelimbs as hydroplanes in the manner of sharks.

It is unclear why an aquatic animal such as *Acanthostega* should have limbs with digits, rather than fins. Sedimentological work shows that *Acanthostega* probably lived in shallow, weed-choked pools and streams. Perhaps a fringe of digits would have helped the animal negotiate such clogged waterways. Perhaps legs with digits were useful for a fish that needed, *in extremis*, to get around in water of negative depth – that is, dry land. Or perhaps, if we insist on telling evolution as a story, we might be forced into thinking of limbs not as fleshy fins perfected for life on land but as specialised fins adapted to get a beached fish back under water as quickly as possible. After all, occasional stranding might be an occupational hazard for a fish habitually living in extremely shallow water.

In short, *Acanthostega* tells us that whatever limbs with digits evolved for, they did not evolve for walking on land. They evolved, for a purpose we can only guess at, while tetrapods still lived very fishy lives, confined to water. The very existence of *Acanthostega* inverts the tale of the brave vertebrate venture as pilloried in Larry Gonick's cartoon. That vertebrates came ashore, and then evolved limbs, is far from a foregone conclusion. On the evidence of *Acanthostega*, what seems to have happened is that tetrapods evolved limbs *before* they came ashore, for reasons unconnected with walking on land.

Once the shock of finding just how fish-like *Acanthostega* really was had worn off, Jenny and Mike had a quick look at *Ichthyostega*. One evening in the early 1990s, I was at a convivial meeting of palaeontologists in the basement bar of the Norfolk Hotel in South Kensington, one of many bolt-holes patronised by researchers at the Natural History Museum. I was sitting at a small, snug table with Jenny. 'Take a look at this', she said, drawing out a large colour print from an inside pocket. The print showed the brown bones of a fossil limb, laid out on a grey background as clearly as a diagram. I counted the bones. I counted the digits. Then I counted them again, this time more carefully. The specimen, Jenny said, was a hindlimb of *Ichthyostega*. It had seven digits.

Work on early tetrapods in the 1980s and 1990s transformed the field by giving palaeontologists a new idea – a mental picture or 'search-image' – of what they should be looking for. The work on *Ichthyostega* and *Acanthostega* gave people a picture of what extremely fish-like tetrapods looked like.

I had also had my own experience with search-images of fossil fishes, although of a kind rather different from the fishes we now call tetrapods.

The Hall of Fossil Fishes did not outlast my childhood. Times change, and my favourite gallery was replaced by other displays. The fossil fishes were all there, of course, but behind the scenes, in the Department of Palaeontology. In 1983, when I was an undergraduate student at the University of Leeds, I returned to the Natural History Museum as a summer intern, to work on fossil fishes in the Department of Palaeontology. Occasionally I'd wander around the collections and open drawers at random, just to see what was inside. Sometimes I'd come across a specimen that had once been an exhibit in the Hall of Fossil Fishes, complete with its display label. I might have seen it last as a child before the gallery was closed down. Now, it was an old friend returned to greet me.

Among my tasks was to sort out a collection of osteolepid fossils collected in northern Scotland, but my job mainly involved sorting and relabelling fossils of another and completely different group of extinct fishes called pteraspids. I was organising them according to ideas presented in the doctoral thesis of a French researcher, Alain Blieck.

Pteraspids were small, flattened fishes mostly no more than a few centimetres long. They had armoured, hydroplane-like fins sticking out of the sides and large, oval plates of bone covering the back and belly. Each species had a distinctive pattern of bony plates that interlocked to make a rigid, box-like 'head shield'. The armour covered only the front ends. The back ends were free, terminating in a scaly tail.

The pteraspid heyday was the Devonian Period, 400 million years ago, and pteraspids have been extinct for hundreds of millions of years. Nothing remotely like them survives today. The first remains to be found were the oval plates that covered the back and belly. These were once thought to have been the internal skeletons – the 'cuttle bone' – of squid. The place of pteraspids among the fishes was discovered only later on, when enough specimens had been found to make their connection recognisable. A close study revealed that pteraspids lacked jaws. This suggested a relationship with modern lampreys and hagfishes, which are similarly jawless. The absence of jaws, though, need not say anything definitive about relationships; earthworms also lack jaws but this doesn't link them with lampreys in particular. More significantly, the arrangement of structures inside the heads of pteraspids was found to have much in common with equivalent features in lampreys. The way the brain is situated inside the skull; the way the gills connect to the outside; the arrangement of a distinctive nostril-like canal that links the outside to the base of the brain, and so on – all these details of pteraspid anatomy demonstrate a close link with lampreys, even though lampreys lack the distinctive head armour of pteraspids.[26]

Because of their inferred relationship to lampreys and hagfishes, the simplest living vertebrates, the pteraspids were banished to the farthest end of the fossil fish collections, with all the other archaic and ancient things. The technicians set up a desk for me in the collections so I wouldn't have to spend time traipsing back and forth to my little office. There I was, day after day, in my remote spot, with my anglepoise lamp and microscope, my view of the traffic in Cromwell Road thundering past outside, with Alain Blieck's doctoral thesis on pteraspid classification (in French), some supporting papers by a man called Zych (in Polish, but I only looked at the pictures), and a pile of fossils that were last alive in the streams and pools of the long-vanished Old Red Sandstone Continent, almost 400 million years ago.

Most of the time, my job was less a problem of biology than of pattern recognition. All the different head-shield patterns seen in the various kinds of pteraspid were laid out by Blieck in clear diagrams, like so many finches in a bird-watcher's field guide. My task was to place the fragments in the collection in the right species, according to Blieck's pictures. In a few days I had mastered the subtle differences in shape and line between the dorsal plate of this one, the ventral plate of that, and the different way that the armoured fins or 'cornua' used to flare in each kind. After a while the patterns became etched on my mind's eye. If I had taken my experience out to fossil exposures from the Old Red Sandstone Continent, my pteraspid 'search-image' would have been as finely tuned as Gabriel's was for hominid teeth in the western Turkana.

This task had very little to do with what the fishes were like as living animals. All I had were fragments, which I could link to larger and more certainly known fragments that were sufficiently informative to have a name. I might as well have been doing the same thing with stamps or cigarette cards. The relationships that these fishes had with living animals is so distant that any attempt to clothe them in flesh, to make them swim, requires a leap of faith.

However, this leap must in some degree be fuelled by comparison with the animals that live around us today. If this were not possible, we would not be able to make any sense of fossils at all. When all that was known of pteraspids were their back and belly plates, palaeontologists compared pteraspids with squid. When more specimens were found, revealing more complete animals with scaly bodies and tails, people realised that pteraspids were more like fishes than squid. Further, more detailed examination of pteraspid head anatomy, afforded by still better fossils, allowed researchers to propose an alliance between pteraspids and lampreys.

When we look at pteraspids now, we interpret them in terms of lampreys; that is how they 'make sense' to us. But the model of a pteraspid in terms of a lamprey is as provisional as that which once linked pteraspids with squid. The lamprey model retains its validity today not because it is true but because it describes the evidence more concisely than any other hypothesis so far advanced. This, in itself, does not close the door to any future hypothesis that might explain the anatomy of pteraspids in a new and different light.

After I graduated from the University of Leeds in 1984, I went to study for my doctorate at the University of Cambridge. I was three doors down from Jenny Clack's office, and I shared a lab with her graduate student, Per Ahlberg. Per was working on lobe-finned fishes, and was a member of the successful 1987 expedition to Greenland. This gave Per extensive first-hand knowledge of *Acanthostega*. After he got his PhD he left Cambridge for a postdoctoral position in Oxford. He took his knowledge of *Acanthostega* with him: this amounted to a totally new mental tool-kit of search-images for early and extremely primitive tetrapods. With *Acanthostega* in his mind's eye, he ferreted around after fossil fishes in museum collections. He kept coming across pieces of fossil jaw labelled as fishes, but which in fact came from hitherto unknown forms of tetrapod. The curators who had originally identified the specimens as fishes, often decades before, were not negligent or stupid – they were doing the best they could with what was known at the time, and they did not have the benefit of Per's new *Acanthostega* search-images. Without such information, earlier curators could have had no idea what a really primitive tetrapod ought to have looked like;

with no suitable mental pictures, all they saw were fishes. Per came to these collections not just with fresh eyes, but eyes conditioned by exposure to the latest tetrapod discoveries. Where other people saw anonymous jaw fragments from lobe-finned fishes, Per saw tetrapods as if they were beacons on hilltops.

Per tracked a few specimens of these tetrapods back to their source, a place called Scat Craig near Elgin, in northern Scotland, where a stream cutting had exposed Late Devonian rocks. Piecing together his finds from Scat Craig unearthed both in museum collections as well as in the field, Per found just enough to put together a tetrapod even more fish-like than *Acanthostega*. This creature, *Elginerpeton*, had limbs that would have borne hands and feet, although Per has yet to recover material as spectacularly demonstrative as that of *Acanthostega*. We do not know how many digits *Elginerpeton* wore on its limbs. It could have been three, nine, six, five, or some other number. Whatever it turns out to be, there is no longer any reason to think that it *must* have been five.[27]

The valuable mental pictures of the earliest tetrapods, made possible by the work on *Acanthostega*, gave people some idea of what the immediate ancestor of tetrapods looked like. As long ago as 1938, a British researcher, Stanley Westoll, wrote a short report in *Nature* on a skull roof of a Devonian lobe-finned fish called *Elpistostege* which he claimed was a tetrapod.[28] The specimen was suggestive, but with no definitive evidence to mark *Elpistostege* as a tetrapod, the status of *Elpistostege* remained undecided; all Westoll had was a skull, with no evidence of limbs. Decades later, Per Ahlberg came back to the same problem that had confronted Westoll. Unlike Westoll, however, he was armed with the fruits of decades of discoveries, and knowledge of what extremely primitive tetrapods looked like – knowledge not accessible to Westoll. Per realized, along with several other workers in Europe and North America, that *Elpistostege* was not, in fact, a tetrapod, although it was not far off being one. It belonged to a small group of lobe-finned fishes, now called the elpistostegids and thought to be the closest relatives of tetrapods among all the other lobe-finned fishes.

If *Acanthostega* is a lobe-fin with legs instead of fins, elpistostegids are the converse: tetrapod-like animals with lobe-fins instead of legs.

Elpistostegids lack all the unpaired fins characteristic of fishes, such as the anal and dorsal fins, retaining only the tail and the paired, lobed, pelvic and pectoral fins. Again, unlike most fishes, the bodies of elpistostegids are flattened from top to bottom, like alligators, rather than side to side, like typical fishes. Elpistostegids are altogether much more what we'd expect the tetrapod ancestor to look like, much closer, indeed, than *Eusthenopteron*.

Elpistostege is no more a missing link than is *Eusthenopteron*. In cladistic terms, however, elpistostegids stand close to the node among the osteolepiforms that represents the latest common ancestor of tetrapods – an ancestor which we can never discover, not even in principle. Just when we think we will have found this ancestor, it will surprise us with a welter of peculiarities such as internal gills or twelve-digit limbs.

A recent cladogram (Figure 8) of lobe-finned fishes and tetrapods, drawn up by Per Ahlberg and Zerina Johanson of the Australian Museum in Sydney, says more about the emergence of tetrapods than any amount of storytelling.[29]

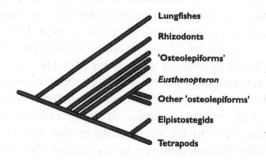

Lungfishes

Rhizodonts

'Osteolepiforms'

Eusthenopteron

Other 'osteolepiforms'

Elpistostegids

Tetrapods

Figure 8. A cladogram showing the pattern of relationships between tetrapods and other lobe-finned fishes.

This cladogram shows, first, why the osteolepiforms have been hard to define as a discrete group. The reason is that osteolepiforms, as a group, are so inclusive. As well as including *Eusthenopteron* and a lot of other, rather anonymous-looking fossil lobe-fins, osteolepiforms also include groups as diverse and distinct as elpistostegids and tetrapods. This means that osteolepiforms cannot be defined uniquely, on their own terms, once these other groups are removed from among them. In other words, there is no simple prescription of features that will uniquely define an osteolepiform, without having to include elpistostegids and tetrapods in the recipe as well. The emphasis on one particular osteolepiform, *Eusthenopteron*, as something close to a tetrapod ancestor has downplayed the fact that nobody has managed to find a set of features that defines osteolepiforms uniquely, as a group, without also including tetrapods in the same group.

But if osteolepiforms cannot be defined as a group, how can palaeontologists recognise them at all? The answer is that osteolepiforms represent less a group than a grade of organisation. A basic osteolepiform looks, in a general way, like the common ancestor from which elpistostegids and tetrapods evolved. In other words, osteolepiforms represent a pool from which certain forms, such as tetrapods, evolved specialisations that took their general level of organisation qualitatively beyond the osteolepiform grade. Osteolepiforms can be defined only by what they lack; they are limbless tetrapods, they are what is left once tetrapods have been taken away.

Another specialised group of lobe-finned fishes was the rhizodonts. These were large animals – the size of alligators – with digit-like structures within their fins.[30] What does this mean? It could be that rhizodonts were 'failed evolutionary experiments' on the way from basic lobe-fins to tetrapods. Could rhizodonts have been the 'missing links' between elpistostegids – which, apart from their lack of digits, were very similar to tetrapods – and tetrapods themselves? Ahlberg and Johanson's cladogram shows that the common ancestor of rhizodonts and tetrapods was a generalised lobe-fin, similar to an osteolepiform. This probably did not have digits, so the digit-like structures in rhizodonts and tetrapods evolved quite independently of one another, possibly for different purposes. The existence of digit-like structures in the otherwise very fish-like rhizodonts shows that limbs with digits were not inevitable acquisitions of a bold march onto land in just one group of lobe-fins, but appeared independently in at least two groups, perhaps in connection with a specialised aquatic lifestyle.

We labour under a burden of hindsight. Were we able to go back to the world of *Acanthostega*, rhizodonts and elpistostegids, and to put aside our knowledge of the subsequent fate of tetrapods – their subsequent evolution into amphibians, reptiles, birds and mammals, including ourselves – we might attach no special import-ance to the digits of *Acanthostega* over the digit-like structures in rhizodonts, or even the lack of digit-like structures in elpistostegids. Tetrapods would have been just another kind of fish.

In the end, we are always faced with the mute puzzle of the fossil itself, a fossil that may fall, in its form, outside the range of anything we have seen, or of anything which exists in the present-day fauna of the planet. The present-day range of non-tetrapod lobe-fins is very sparse compared with what once existed. There are no more rhizodonts or elpistostegids, and no present-day lungfish or coela-canth looks much like *Elpistostege* or *Acanthostega*. Today, there are no fishes with legs and digits, and no tetrapods with seven or eight digits per limb. However, we are forced to interpret fossils in terms of the familiar forms alive today and these might look very different from the fossils that confront us. Given this difficulty, it is best not to make the act of interpretation harder by expecting fossils to con-form to preconceived stories about the course that evolution may have taken. As far as we are able, we should try to see fossils as they are, not as we want them to be.

When we come across the fossil of a hitherto unknown form of life, we range it against mental pictures – search-images – of the organisms we know, in a kind of identity parade. This is how Jarvik, over the course of decades, came to see *Ichthyostega* in terms of his experience with fossil fishes such as *Eusthenopteron*. We isolate features of the new form that seem recognisable, because they match features in the things we know. Per Ahlberg, Jenny Clack and Mike Coates, looking at new fossils in the light of Jarvik's legacy, could pick out specimens of tetrapod bones in collections labelled as fishes and long believed to be nothing out of the ordinary. In this way, we come to describe new forms in terms of their greater or lesser resemblance to familiar things.

Long before I first entered the Hall of Fossil Fishes, my parents took me to see the life-sized dinosaur sculptures in Crystal Palace Park in Sydenham, a district in south-east London. The Crystal Palace was a huge glass exhibition hall, the centrepiece for the Great Exhibition of 1851. It was originally put up in Hyde Park in central London, but was later moved to Sydenham, where it stood until destroyed by fire in 1936. The Palace gave its name to the park (not to mention a soccer team), even though only the foundations remain. The Palace may have gone, but the concrete dinosaurs, also created for the exhibition, are still there, poised on an island in an artificial lake.

The dinosaurs, created by the sculptor Benjamin Waterhouse Hawkins, were a hit at the Great Exhibition. Then, as now, the public was gripped by dinomania. In 1842, Britain's leading anatomist, Richard Owen, had created the term *Dinosauria* – the 'terrible lizards' – to refer to several unusual fossils that had been unearthed in England. One of these fossils had been named *Megalosaurus* ('enormous lizard') and another *Iguanodon* ('iguana tooth'). These fossils seemed to have belonged to reptiles of great size, like lizards but much larger – and yet somehow different, the way that unicorns are different from horses, in kind as well as degree. These creatures had a lightness, something of the bird about them; they had a poise quite unlike the saurian, more like that of a mammal.

Owen deserved his reputation as an anatomist; 150 years later, his insight still seems remarkable. He took these broken fragments and grouped them together as members of a wholly distinct order of creation, akin to reptiles but with something of the appearance of birds and mammals, and yet different from all three. The word 'dinosaur' was more than just a word – it was a concept, an admission that there once existed creatures that did not fit into the categories of creature alive today – not reptile, not bird, but something else, something not previously seen.

Today, Hawkins's restorations seem quaint. *Iguanodon* is a bruiser of a beast, four feet squarely on the ground like a hippo, or sprawling like an outsize monitor lizard, a small horn on its nose like a dollop of rice pudding. *Megalosaurus* looks like something out of Egyptian myth, with the body of a lion and the head of an alligator.

In truth, Hawkins based his sculptures on extremely fragmentary material, extrapolating from the little that 1850s science knew about dinosaurs.

The problem for Hawkins was that dinosaurs were completely extinct, and there was nothing quite like them around that a sculptor could use as a model. Dinosaurs were something like lizards and alligators, only inflated beyond a size one would expect a modern reptile to attain. Alligators and lizards could be used as starting-points. But dinosaurs did not sprawl in the same way as reptiles, and their limbs, for all their great size, seemed surprisingly delicate, more like those of birds.

The confusion is evident in Hawkins's sculptures, which are less like re-creations of real dinosaurs than a compound of models mixed together with a small sprinkling of real evidence. The result is a magnificent parade of mythical beasts almost entirely of human imagining, like gryphons or Borges's unicorn. (The effect, though, is undeniably impressive. When I revisited Crystal Palace Park after an absence of thirty years, the dinosaurs still looked as lithe and lifelike as I remembered. In the slanting rays of the evening Sun, they really did look like visitors from out of Deep Time.)

Nowadays, we would restore both *Iguanodon* and *Megalosaurus* not as quadrupeds, but in a bipedal, somewhat bird-like poise, with heads out in front, tails held straight out behind as counterweights. These restorations are based on more complete fossils, including complete skeletons, and a century and a half of knowledge that was unavailable to Hawkins. But we patronise Hawkins at our peril. He was a pioneer, doing the best with the limited knowledge he had. For him, the great age of dinosaur discovery was still to come. Knowledge about dinosaurs increased at a great rate, especially in North America, when palaeontologists started discovering dinosaurs in the West. Most of the best-known dinosaurs – *Triceratops*, *Stegosaurus*, *Allosaurus*, *Tyrannosaurus* – come from bones excavated in the western United States, and owe their discovery to two men, Othniel Charles Marsh of Yale, and Edward Drinker Cope of Harvard, whose energy, rivalry and mutual hatred led to ever greater frenzies of collection, as each strove to outdo the other.[31]

When dinosaurs were first presented to the Victorian public, many were seen as agile, fast-moving creatures, more like large flightless birds than slow-moving reptiles. But the tide slowly turned against this idea. By the time I was given my first dinosaur book, in the 1960s, dinosaurs were seen as slow and sluggish. In those days – not really so very long ago – a dinosaur such as *Iguanodon* was less a sprightly ostrich than a sick kangaroo, standing upright with tail dragging along the ground. Large, long-necked dinosaurs, such as *Diplodocus* and *Brachiosaurus*, were thought to have been too big and heavy to have supported their own weight on land, and were often pictured floating around in lakes.

These views changed again in the 1970s, when a dinosaur 'renaissance' painted the dinosaurs once more as active, intelligent creatures. A modern child, growing up now, would have no cause to imagine that dinosaurs were seen in any other way, yet fashions have changed, and have changed back. Our ideas about dinosaurs could change yet again, once more evidence becomes available.

To interpret fossils, we need to be armed with search-images. Fossils may take forms beyond our imagination because our imagination is shaped by the things around us; we are constrained to make sense of fossils in the light of the things we know, we cannot do otherwise. Because search-images are subjective, based on personal experience, they are far from the objective tests that science would demand in an ideal world; they are necessary evils. However, the cause is far from hopeless. Prior experience and new knowledge reinforce each other to broaden our horizons. For example, the work on lobe-finned fishes by Säve-Söderbergh, Jarvik and others guided the search for early tetrapods. Once discovered, tetrapods such as *Acanthostega* illuminated and enriched, in turn, our understanding of lobe-finned fishes. In this way we describe new and unusual forms in terms of their greater or lesser resemblance to familiar things. The story of palaeontologists struggling, with inadequate search-images, to make sense of things that are new and strange – and having their understanding dashed or strengthened by subsequent evidence – is a tale that can be endlessly repeated. Here are just a few more such stories.

Conodonts are tiny, tooth-like fossils that geologists have studied for more than a century. They are found in rocks of the Cambrian

Period (around 543 to 505 million years ago) up to the Triassic (248 to 213 million years ago), often in huge quantities. Sometimes a certain kind of conodont will be found in a particular rock stratum so densely that geologists can recognise the stratum from its conodont contents alone.

Some conodonts are simple cones. Others are spiny and complicated, articulating to form apparatuses of unknown purpose. For decades, no evidence existed that might bear on what kind of creatures conodonts came from. Imagine the job facing palaeontologists of the future, obliged to reconstruct human history from billions of sets of discarded dentures, and these alone. Despite the total lack of evidence, conodonts have been associated with all kinds of peculiar fossils. Some wondered whether conodonts were parts of animals at all, or could have been crystals that formed inside plant tissues.

Rapid progress was made following the discovery of 350-million-year-old fossils from Scotland, in which conodonts were consistently arranged at the front ends of the delicate, faint impressions of thin, worm or eel-like animals. More and similar fossils were found in South Africa. Apart from the conodonts themselves, the animals to which they were attached were soft-bodied and unlikely to be fossilised in the normal course of events. Although conodonts are hard and phosphatic, and routinely preserved in great numbers, it takes a special kind of sedimentation (usually under conditions of exremely low ambient oxygen) to preserve the organic remains of soft-bodied creatures such as conodont animals.

Later work on exquisitely preserved specimens has revealed that conodont animals have a distinctly vertebrate-like anatomy. The vertebrates (the group that includes you, me, Fred the cat, all other tetrapods and all the fishes) have a very distinctive anatomy. The head is usually clearly demarcated from the body, has a large brain in a protective box – the skull – and prominent organs of special sense, such as eyes and ears. The vertebrate trunk and tail are supported by the vertebral column, which forms the purchase for a series of segmental, V-shaped muscle blocks – a fact that you can verify instantly the next time you see a side of smoked salmon in your delicatessen. Like a vertebrate, a conodont animal has a trunk and tail divided into segmented muscle blocks, and a large pair of eyes at the front.

Among the most distinctive features of vertebrates are their hard tissues, bones and teeth. Bone is a substance unique to vertebrates, and is made of a crystalline mineral – hydroxyapatite, a form of calcium phosphate – set in a scaffold of organic, collagenous tissue. Teeth are made of enamel, another tissue based on calcium phosphate and the hardest substance produced by any living organism, supported in turn on dentine (another calcium phosphate-based material) and bone. The conodonts themselves (the term 'conodont' is properly reserved for the tooth-like structures, not the animals to which they belonged) are made of the same substances, bone and enamel, that one finds in vertebrate teeth, including yours and mine.

Most palaeontologists now agree that conodont animals are most closely allied to the vertebrates – backboned animals, including ourselves. At present, nobody really knows precisely where they fit into the vertebrate cladogram. Our uncertainty is simply a reflection of the lack of suitable models. Although conodonts resemble vertebrate teeth in a general way, they are nothing like anything found in any living vertebrate. Because of this uncertainty, you cannot assume without question that they were used for the same purposes as teeth are used for today, such as biting and grasping.[32]

Conodont animals were once hugely abundant, but they perished, leaving only the strange, tooth-like conodonts, like so many geological Cheshire cats leaving their smiles to posterity. As long as the true identity of conodont animals remained an enigma, people could afford to ignore the question of where they might have belonged in the history of life. Now that their status as vertebrates is generally accepted, we must take account of their contribution to vertebrate evolution. For if conodont animals were akin to the earliest vertebrates, they may have been vastly more numerous than the pteraspids, sarcopterygians, and the other amazing creatures I encountered as a boy in the Hall of Fossil Fishes. The story of the conodont animals, then, shows us that the diversity of fossil forms is likely to be greater than the diversity of creatures alive today. This makes choice of suitable models sometimes fraught, but can open entirely new, unexpected views on the history of life.

Our experience of living vertebrates suggests a limited number of forms, of 'body plans'. But the existence of creatures like pteraspids and *Acanthostega*, both quite unlike living forms, suggests that many more kinds of vertebrate body plan existed than we see around us in the extant fauna. Conodont animals, like pteraspids and *Acanthostega*, were vertebrates and, like them, different in kind from any extant form. No modern vertebrate combines the sinuous, thread-like form with the complex basket of pointed, phosphatic teeth that we see in conodont animals. The teeth themselves have a microstructure generally similar to that of vertebrate teeth, but in some ways rather different. This is itself a source of debate between those who would see conodont animals as vertebrates and those who would not, but I see a third way: once upon a time, there were many more kinds of bone, dentine and enamel structure than there are now. Students of fossil fishes have built up a lengthy catalogue of unusual bone tissue-types seen in fossil fishes and not in modern vertebrates. The particular microstructure of conodonts is simply another addition to that list. If conodont animals are fish-like vertebrates, they represent a whole new strain of vertebrate evolution, and one which we find hard to place in the scheme of vertebrate evolution – armed, as we are, only with our limited experience of the forms of bone seen in animals alive today.

Nonetheless, the arrival of conodonts in the realm of vertebrates changes everything we thought we knew about the early history of vertebrates. The usual picture of the first vertebrates is of sluggish, armoured creatures, like the Devonian pteraspids I studied as a student. Fossils of similar armoured, jawless forms are known, in rare and fragmentary form, back as far as the end of the Cambrian (543 to 505 million years ago). If armoured jawless animals were representative of early vertebrates – if that's all there were – we'd have a picture of a slow, unsteady start at the end of the Cambrian, picking up through the succeeding Ordovician (505 to 438 million years ago) and Silurian (438 to 408 million years ago) to a blossoming in the Devonian, the traditional 'Age of Fishes'.

But if conodont animals are admitted to the vertebrate aquarium, they were there right at the beginning, in the Cambrian, and in large numbers. Compared with the profusion of conodonts, armoured creatures such as pteraspids would have been relatively unimportant in the range of vertebrate form.

Suddenly the carefully drawn story of early vertebrate evolution as I learned it in the Hall of Fossil Fishes is turned upside down. Do all those armoured, jawless animals – groups like the pteraspids – become side-branches of evolution? Could later vertebrates have had instead a common ancestor among the conodont animals? Or do conodont animals represent their own, unique flourish of early vertebrate diversity? And how diverse were they? At present we have evidence for only a handful of conodont animals, and these all seem to have been long, thin and eel-like. But there are hundreds of conodont species, each named according to the teeth-like conodonts alone, with no knowledge of what kind of animal they belonged to. We do not know what the animals attached to these conodonts looked like. Just because the conodont animals that have been found so far look long and slinky, this need not apply to all conodont animals.

Whatever the answers, conodont animals – known only as fossils – have changed the way we look at the past. The familiar past, based on the ranked fossils in the old Hall of Fossil Fishes of my boyhood, was in reality a much stranger place. The Hall of Fossil Fishes might have presented a truer picture of the Age of Fishes had the bulk of its space been devoted to conodonts, with just one small case, in a dark corner, devoted to pteraspids, sarcopterygians and their relatives.

Conodont animals shared the seas with even stranger things. Some, anchored to the sea floor by long stalks, or perhaps slowly crawling around, looked at first sight like tethered starfishes. They had platy skeletons of calcite, just as starfishes have today. Calcite is a form of calcium carbonate, distinct from the calcium phosphate of conodonts, vertebrate bones and teeth. A skeleton of calcite plates – each made from a single calcite crystal – is a sure sign that an animal is an echinoderm, a member of the group of 'spiny-skinned' animals that includes starfishes, brittle-stars, sea-cucumbers, sea-urchins and the less familiar deep-water sea-lilies.

Echinoderms today have a number of distinctive features that set them well apart from anything else in the animal world. Apart from the skeleton made of calcite plates, their bodies have a radial symmetry based on the number five. Many starfishes have five arms, for example. They also have a multi-purpose arrangement of hydraulically driven, inflatable 'tube feet'. An individual echinoderm will have hundreds of tube feet, arranged in rows or 'ambulacra', and they are used for everything from moving the animal about, to collecting food. Like the rest of the body, the ambulacra have a five-way arrangement. Each starfish arm carries one ambulacrum. This system of ambulacra and tube-feet is unique to echinoderms.

Echinoderms have been around since the Cambrian Period. In all that time, a wide variety of echinoderm body shapes has come and gone. Today, there are five basic sorts, or 'classes', of echinoderm; back in the Cambrian, and the Ordovician Period that came immediately after, there were perhaps as many as twenty classes of echinoderm, each as distinct from the other as, say, starfishes are distinct from sea-urchins. In those days, echinoderms were far more diverse than they are now. There were forms with stalks or without them, forms with ambulacra held on arms, others with ambulacra spiralling round a basketball-shaped body; even tunnelling forms shaped like screw-threads, boring their way through the sea floor. Some of these echinoderms did not have five ambulacra, but three, two, one – or none. Some echinoderms lacked the characteristic five-way symmetry. In fact, a few had no recognisable symmetry at all.

A look at the diversity of ancient echinoderms shows that the invariable pattern of symmetry based on the number five is true only for modern echinoderms. Some ancient forms were so unlike modern echinoderms that only the shared possession of a calcite skeleton can be used to link them together at all. The fossil *Cothurnocystis*, for example, from the Ordovician of Scotland, has a calcite skeleton, but neither symmetry nor ambulacra. Apart from its calcite skeleton, nothing about *Cothurnocystis* says that it is an echinoderm.

For most palaeontologists, the fact that *Cothurnocystis* has a calcite skeleton is enough to make it an echinoderm, without question. The calcite skeleton is a 'key feature', a kind of rubicon. Once an animal has a calcite skeleton, it is an echinoderm, and that is that. No other solution is possible.

Echinoderms did not arrive fully formed, complete with all the distinctive features by which we would recognise them today. The traits that make echinoderms what they are appeared one by one – and, one by one, they can be lost. Some modern echinoderms, such as sea-cucumbers, have all but lost their calcite skeletons, and yet for all their present *déshabillé*, nobody denies that they are echinoderms. Could other echinoderms, in the past, have lost their skeletons? And if they did, how would we recognise the hidden echinoderm ancestries of these animals? Without ambulacra, five-way symmetry and, most of all, without calcite skeletons, there would be nothing to link them in particular with echinoderms.

It has been known for a long time that vertebrates and echinoderms share a common ancestry that excludes the ancestry of other great animal groups, such as insects or molluscs. To be precise, echinoderms are related to the chordates, a group that includes vertebrates, as well as some lesser-known animals, the tunicates (or 'sea-squirts') and the lancelets – small, darting creatures of inshore waters that look vaguely fish-like, rather like animated anchovy fillets. Chordates are distinctive in that their bodies are supported, at least during some part of their lives, by a firm yet flexible rod of collagenous material, the 'notochord', that runs from the head to the tip of the tail. Lancelets retain a notochord throughout life. Tunicates have notochords only as larvae. In vertebrates, the notochord forms a kind of scaffolding for the bony vertebral discs, which replace and supplant it. You, me, Fred the cat, all other tetrapods and all fishes are chordates as well as vertebrates.

Among many other distinctive features, chordates have a highly modified throat region, or pharynx, which is perforated by a series of hole or slit-like openings called 'pharyngeal slits' that connect the front part of the gut (behind the mouth) with the outside world. In fishes, these openings become the gill slits. In modern tetrapods, these gill slits are suppressed during development, but in sea-squirts and lancelets the slits have become elaborated into a complex structure for filtering particles of food from seawater.

Modern echinoderms have none of these things, yet chordates are closer cousins to starfishes than to squid or lobsters. What did the most recent common ancestor of echinoderms and chordates look like? Looking at me and my cat Fred, you can sketch the essential features that our most recent common ancestor would have had; in the case of Fred and myself, it would have been a mammal. Answering the question is much harder in the case of, say, myself and a starfish, because chordates and echinoderms have had a much longer time to acquire their own special features, a process of acquisition that has all but obliterated any family resemblance. The only clues come from very early in the embryonic life of a young echinoderm or chordate, when they are hardly more than balls of cells. We know what our common ancestor looked like as a microscopic embryo – but all clues to the adult appearance of the most recent common ancestor of echinoderms and chordates have been erased by time and change. It is usually assumed that echinoderms acquired their platy skeletons, and chordates – such as myself, Fred, sea-squirts and lancelets – acquired their notochords, perforated throat regions and so on, separately, long after the divergence between echinoderms and chordates. One scientist, Richard Jefferies of the Natural History Museum, exposes this assumption as just that – an assumption.

For four decades, Jefferies has been looking at echinoderm-like fossils such as *Cothurnocystis*, from the Ordovician of Scotland, and has been seeing vestiges of what might be a closer connection between chordates and echinoderms. Although *Cothurnocystis* has the calcite plates of an echinoderm, a closer look reveals a long 'tail' that could have held a notochord. The blobby, asymmetrical body of *Cothurnocystis* has a row of openings that look rather like the pharyngeal slits of the kind one would expect to see in chordates, such as fishes. *Cothurnocystis* could have been a chordate in echinoderm clothing.

Cothurnocystis is remarkable in that it shows, in the same animal, features which today are used to identify two distinct groups. It has a calcite skeleton, just like a modern echinoderm, but it also has, in Jefferies' view, a notochord-supported tail and pharyngeal slits, just like a modern chordate. *Cothurnocystis* is both a chordate and an echinoderm – and it is neither. Nor is it a 'missing link' between the two. Rather, it could represent its own distinct form of creature, related to both echinoderms and to chordates, but forming some part of the story of chordate and echinoderm evolution that is irrecoverably lost, at least given the range of modern vertebrates and echinoderms. Imagine that tomorrow, some deep-sea submersible recovers live specimens of *Cothurnocystis*, living at the bottom of the Pacific Ocean. Their internal structure proves to be as Jefferies predicted, with a notochord and a tunicate-like filter-feeding pharynx. How would we classify it? Echinoderm? Chordate?

Jefferies looks at the same fossil history as everyone else, but reads into it a totally different story. Rather than chordates and echinoderms having a common ancestor of unknown form, he sees our common ancestor as looking something like *Cothurnocystis*. In Jefferies' view, the story of chordate and echinoderm evolution is not one of independent divergence and subsequent diversification, such that chordates gained the notochord and pharyngeal slits, while echinoderms separately gained calcite skeletons and tube-feet and so on. Rather, Jefferies' account is one of loss. Starting with an animal similar to *Cothurnocystis* – an animal in which is combined chordate features such as a notochord-supported tail and pharyngeal slits, together with echinoderm features such as a calcite skeleton – echinoderms lost their pharyngeal slits and tails, only later to gain symmetry and ambulacra, while chordates lost their calcite skeletons. Some chordates, the vertebrates, later acquired skeletons of bone – calcium phosphate, rather than calcium carbonate.

All Jefferies has really done is choose to apply an unconventional model to a fossil. Most people would see the calcite plates of *Cothurnocystis* and automatically assume that it must have been an echinoderm. On that assumption, they then 'read' the rest of the animal's anatomy in the light of modern echinoderms. But Jefferies questioned that assumption, showing that there was little to link *Cothurnocystis* with modern echinoderms rather than chordates. In choosing to apply a chordate 'model' to *Cothurnocystis*, Jefferies could interpret its anatomy – and chordate history – in an entirely new way. Jefferies has yet to win over many people to his way of thinking and it is easy to see why. With an animal as strange as

Cothurnocystis it is hard to justify the use of one modern model over another in its interpretation.

But let us suppose for a moment that Jefferies is right. If he is, then the chordates – and ourselves among them – could be a lost class of 'naked' echinoderms, the echinoderms that shed their skeletons like butterflies emerging from cocoons. We could be the starfishes that lost their calcite skeletons.[33]

The ancient seas were home to stranger animals still. Here is a rather fanciful description of one of them: 'Behind the oversized head . . . stretched a long, smooth green horizontal thorax. Bifurcated pale pink tentacles . . . rose in a crest along the back . . . A thick uplifted tail ended with a purple bugle flare. Perhaps strangest of all, seven pairs of lower "legs" or supports lined both sides of the body, not legs or limbs in the traditional sense but *poles* or long sharp-tipped spikes . . .'

Well, actually, I cheated. This is not a description of a real animal that once existed, but an alien from a science-fiction novel, *Eternity*, by American author Greg Bear.[34] But Bear was inspired by a real fossil of an inch-long animal from the Cambrian Burgess Shales of Canada, described in the scientific literature by Simon Conway Morris of the University of Cambridge. Conway Morris named it *Hallucigenia*, on account of its bizarre, 'dreamlike' appearance.[35] The implication was that you would see things like this only in psychedelic fantasies or, as Greg Bear imagined, on another planet.

Here was a creature so different from anything known that it bore no comparisons whatsoever. A fossil that has no model of any kind is, in a sense, a hopeless case. Without a contemporary model to provide a search-image, its place in the wider scheme of things remains unknown. All it does is remind us that Deep Time can throw up fossils that puncture the conventional idea that the present is a key to the past, that the diversity we see today is all that is possible.

Further work on some of the specimens of *Hallucigenia* showed that Conway Morris had, in fact, interpreted *Hallucigenia* upside down. The seven pairs of unjointed spikes were not stilts, but long spines carried on the animal's back. The tentacles, originally placed on the back, were paired, and were reinterpreted as legs, underneath the animal rather than on top of it. We should not laugh – like

Waterhouse Hawkins and his dinosaurs, Conway Morris was doing the best he could with the evidence at hand. Once *Hallucigenia* had been turned the right way up, it could be interpreted in the light of modern models.

The models came, ultimately, from animals called velvet worms, which live under logs in tropical forests. Neither worms nor made of velvet, they are cousins of the great group of jointed-limbed animals – the arthropods – that includes the crustaceans, spiders and scorpions, centipedes, millipedes and insects. All modern velvet worms are terrestrial, but much recent palaeontological work shows them to be the last in a once-flourishing group of marine animals. *Hallucigenia* is now seen as one of these.

Back in the Cambrian, these relatives of the modern velvet worm formed a diverse and important group of animals, the lobopodians, that included creatures at least as distinctive as *Hallucigenia*. One was *Anomalocaris*, a metre-long predator that looked like a stream-lined lobster with a circular mouth like a kitchen garbage-disposal unit. Another was *Opabinia*, a creature with five stalked eyes, a prawn-like body and a long, hosepipe-like snout tipped with a pair of barbed jaws – less like a real animal than a design for a vacuum cleaner by Salvador Dalí. And yet the only living representative of this group is the humble velvet worm, which lives exclusively on land. This remarkable group of animals has completely vanished from its oceanic home.[36]

Conodonts, *Cothurnocystis*, *Hallucigenia* all lived a very long time ago, in an epoch so remote that it may as well have been mytho-logical. Perhaps it is not really so surprising that creatures of such antiquity seem so strange today. However, some very strange creatures survived almost until the dawn of history. Our ancestors may have met them.

Until almost historical times, in the Americas, lived massive land animals armed with claws like scythes that could have gored a man as easily as swatting a fly. The last ones we know about, from fossil evidence, died out as recently as 11,000 years ago, so undiscovered ones could have survived into more recent times. The first Americans penetrated Alaska at least 13,000 years ago, so human beings could have seen these animals alive.

These creatures were ground-dwelling relatives of the modern sloths of South America. The ground sloths lived in South America, as far south as Patagonia, but also penetrated North America. Their remains have been found at the La Brea tar pits, just off Wilshire Boulevard in modern Los Angeles. Some of these ground sloths were small, others much bigger. *Megatherium* weighed as much as two African bull elephants.

Given that all modern sloths live in trees, and no comparable creature lives on the ground, let alone one that weighs much more than any modern terrestrial mammal, it is hard to 'read' much into the remains of ground sloths. Everything about them seems awkward or contradictory. What did they eat? They had claws like offensive weapons, but their teeth were small and weak. What was their posture at rest? Their vast, bowl-shaped pelvises look as if they were made for sitting, but their massive tails would have got in the way. How did they move? The fore and hind-feet of *Megatherium* were strangely twisted and turned-in, like the bound feet of a Chinese noblewoman, looking acutely painful to the modern eye. And yet fossilised footprints of giant ground sloths show that they managed a respectable speed of about five kilometres an hour.

Giant ground sloths are a mess of riddles, yet Richard A. Fariña and R. E. Blanco, two researchers from Montevideo, Uruguay, have plunged in with a startling suggestion: that *Megatherium* was the largest mammalian meat-eater of all time.[37] The forearms of *Megatherium* suggest that they were designed for sudden bursts of power. Armed with those claws, these arms could have slashed open the belly of an adversary, or overturned a 1,200-kilogramme truck. The researchers suggest that likely prey could have been extinct, jeep-sized, heavily armoured armadillos called glyptodonts. Perhaps a sloth would hook its claws under the armoured rim of a glyptodont carapace, flip it over, and expose the undefended underside to further blows from its claws. Thus opened, the soft interior of the victim could have been chewed by the sloth's unremarkable teeth.

It makes an odd picture – some might say an unlikely one. But it is unlikely only in the light of conventional assumptions about the diverse past derived from the limited selection of animals we have around us today. Nowadays, sloths are harmless animals that live in trees and armadillos are small creatures easily chased under bushes by small dogs. Present-day experience, however, does not rule poss- ible out primeval encounters between megatheria and glyptodonts, however strange these scenes may seem – scenes that could, indeed, have been witnessed by the first Americans.

In all these case-histories, models have been found through which the fossils can be interpreted and placed in the scheme of life. *Cothurnocystis* looks unintelligible at first sight, but nevertheless can be interpreted as either an echinoderm or a chordate. Conodonts and *Hallucigenia* make sense when interpreted as, respectively, vertebrates and marine, armoured velvet worms. Even *Mega- therium*, for all its oddity, is a mammal and a sloth, so there will always be a place to start. And all these animals have one thing in common: their models are all of Earthly origin. If evolution is a fact, it is probably fair to say, as a parsimonious assumption, that all the organisms alive today, and that ever lived, are related to one another. We are all cousins. We can assert this, in the same way that we can assert a common heritage for myself and my cat Fred – by virtue of the many traits that we hold in common.

As far as we can tell, all the organisms that we know share the same fundamental features of biochemistry and metabolism. We all – down to every microbe – have genetic material encoded in long chains of the molecule deoxyribose nucleic acid (DNA) or the very similar ribose nucleic acid (RNA). And no matter which organism you investigate, this genetic information is always accessed by the same elaborate system of protein enzymes and molecules of RNA, and translated into sequences of other chemicals called alpha-amino- acids. These polymerise to form proteins. In virtually every organism one cares to examine, energy is moved around cells in the form of a molecule called adenosine triphosphate (ATP). Ultimately, the chemistry of living things is based on the chemistry of carbon, and the tendency of carbon atoms to link up into long molecular chains. The biochemistry of carbon invariably takes place in aqueous solution: that is, in water. We have yet to find life on Earth based – say – on the chemistry of long-chain germanium molecules stable only in superheated, pressurised sulphur hexafluoride, if such a thing were even possible on thermodynamic grounds.

The fundamental and detailed chemical similarities shared by all creatures on Earth speak strongly in favour of a single common ancestry of all life. In fact, we can paint a fairly detailed picture of the most recent common ancestor of all life, based largely on the similarities listed above: a list that is not exhaustive.

The quest to interpret fossils in terms of modern models rests on the assumption that all life on Earth has a common ancestry, for we can interpret past life only in terms of other living organisms. If this were not possible, we would not recognise the fossils of animals as animals at all. We'd just see them as rocks. This last point is crucial; if we can interpret past life on Earth only in terms of present-day life on Earth, it may be impossible to use these same Earthly models to judge the strength of claims for the discovery of living or fossil organisms elsewhere in the Universe.

In 1996, David S. McKay of the NASA Johnson Space Center in Houston, Texas and his colleagues published a report suggesting, very tentatively, that signs of past life had been found inside a meteorite discovered in Antarctica, but which had originated on the surface of the planet Mars.[38] The meteorite bore the catalogue number ALH 84001.

ALH 84001 contained carbonate minerals. Carbonates, such as chalk and limestone, are commonly associated with life; indeed, the carbonates in ALH 84001 were arranged in globules reminiscent in size and texture of carbonate deposits laid down by bacteria. As well as carbonates, the meteorite contained particles of magnetite, an iron mineral often found as a by-product of bacterial metabolism. Tiny coin-shaped blobs of carbonate with iron-sulphide and oxide rims may represent the 'shadows' of bacteria, in the same way that the brown rings on a table top are reminders of yesterday's parties. ALH 84001 also contains complex carbon-containing substances such as polyaromatic hydrocarbons (PAHs). Compounds as complex as these are usually associated with the metabolic activities of living organisms. Most intriguing of all, ALH 84001 contains tiny blobs and threadlike features which have been compared with bacterial fossils.

All these features could equally well have had an inorganic as an organic origin; carbonates and iron minerals do not always need life to form, and may have been deposited at temperatures of several hundred degrees, inimical to life. Many meteorites contain complex substances such as PAHs, which seem to be found widely in the Universe. The proposed fossils seem much smaller than the smallest Earthly bacteria, and could well be nothing more than mineral inclusions or even contamination inadvertently introduced on Earth after ALH 84001 landed in Antarctica.

These objections are irrelevant. For, like the claims they question, they are based on the unstated assumption that Martian life is similar to Earthly life, and behave in the same way. Carbonates, magnetite, PAHs and so on are all associated with living organisms on Earth, but this need not be true for life on Mars. Even though Earthly bacteria live in all kinds of exotic places as seemingly inhospitable as the surface of Mars – ranging from oil reservoirs miles underground to the sulphurous vents of hot springs – they are still terrestrial creatures made, essentially, of the same things that constitute every other Earthly organism. Earthly bacteria, then, are inappropriate models for Martian ones. Once this is admitted, it is impossible to tell whether the carbonates, iron minerals and so on are or are not the products of life. If the small blobs seen in ALH 84001 really are fossils, how would we know? The only standards of comparison we have are Earthly organisms, which need have no points of similarity with Martian organisms.

J. William Schopf of the University of California, Los Angeles, is an authority on the fossils of bacteria and other microscopic organisms.[39] In his book *Cradle of Life*, he shows how it is possible to recognise signs of life in Earthly rocks, even if the fossils are bizarre in shape. Fossils of microscopic organisms exist that defy interpretation and classification, but they always look like cells of some kind, are always about the right sort of size for cells, lived where other living forms lived, and are made of organic matter. So, it is possible to recognise life of new and unusual kinds. But this need only apply if the fossils concerned – and the models used to interpret them – are of Earthly origin.

In this light, complaints that Martian fossils are too small to be bacteria, or are the wrong sort of shape, and so on, apply only if it is valid to compare Earthly and Martian bacteria directly. If Martian bacteria were made in some fundamentally different way, these complaints might not apply; Martian life-forms could be as big or small or as oddly shaped as you please.

The fact is that we, as Earthly organisms ourselves that know no other form of life, are limited in our ability to make comparisons. We can only compare putative Martian microbes with Earthly ones, because that's all we have. When, in 1976, the *Viking* lander sifted the Martian surface for signs of life and found nothing, the reason could have been because mission scientists (who, in all fairness, were unable to do anything else) assumed that Martian organisms would behave just like Earthly ones. But the Martian surface could have been infested with microscopic Martians and we would never have noticed. When, in 1997, the Mars *Rover* had face-to-face meetings with Martian rocks, we could never know whether the rocks were not equal participants in the exchange.

This discussion about life on Mars may seem extreme, but it serves to expose the limitations of search-images, and draws attention to the fact that we can make sense of fossils only if we have suitable search-images at our disposal. In the case of life on Mars, no search-images are available because the concept of the search-image, inasmuch as it applies to fossils discovered on Earth, rests ultimately on the assumption that all life on Earth has a common origin.

Confronted as we are by the problems and limitations of search-images, we must be pragmatic. Fossils may take forms beyond the capacity of our imagination to fathom, yet we must interpret them against the things we know. Our stock of possible models comes from the range of life-forms that currently exists on the Earth. This range is great, but the diversity of forms that has come and gone over the course of Deep Time is greater. Our choice of model is therefore contingent on what happens to be around right now. Our choice also depends on our individual experience, as we build up our own personal repertoires of search-images. By this halting, inadequate mechanism, we slowly make sense of new forms from the history of life in terms of their resemblance to familiar things. But new knowledge and personal experience reinforce one another and interact to push back the frontiers of ignorance.

In the end, we never see fossils as they are, but only imperfectly, in the light of models that are more or less approximate. Given this constraint, it is surely hard enough to make progress understanding the evidence we have, without leaping way beyond it, with presuppositions about chains of ancestry and descent, and missing links. Such presuppositions are exposed as vacuous once the evidence finally catches up. The tale of the water–land transition, for example, started with the assumption that limbs with digits evolved for walking on land: this was their purpose. The tale ended with the exact opposite: limbs with digits evolved in animals that were obligately aquatic, so they presumably evolved for some other reason.

Should you go hunting for the unicorns that prowl the endless dark corridors of Deep Time, I offer the following advice. Before speculating about the function of a unicorn's horn, take the time to understand the place the unicorn occupies in the pattern of the history of life, so you will be on firmer ground when trying to understand how the unicorn evolved to be the way it is, and the forces that shaped its evolution.

It could be, of course, that the unicorn's horn has a variety of uses, including sexual display, territorial combat, and removing stones from the hooves of other unicorns. Such speculation is fine – provided that we know that the horn has not been glued there; that we already have some idea of the structure of the horn and how it might correspond with similar structures in other animals; and that we have some sense of the unicorn's place in nature.

3 There Are More Things

An armchair presupposes the human body, its joints and limbs; a pair of scissors, the act of cutting. What can be said of a lamp or a car? The savage cannot comprehend the missionary's Bible; the passenger does not see the same rigging as the sailors. If we really saw the world, maybe we would understand it.

Jorge Luis Borges, *There Are More Things* (*To the memory of H. P. Lovecraft*)

The stories we tell ourselves about evolutionary history – such as the tale of how fishes got their legs, as discussed in Chapter 2 – are true only inasmuch as they reinforce our prejudices; they tell us what we want to hear, not what really happened. How do such stories have the potential to be so wrong? The reason is that they become detached from science or, rather, from the limited capacity of science to examine these stories in terms of hypotheses or experiments which can be tested.

Ultimately, the fault lies with our failure to recognise Deep Time as an unsuitable medium for telling stories. When we create a story about evolution, as I showed in Chapter 2, we are constrained to use what we know, using our stock of search-images based, ultimately, on present-day experience. For example, our experience of tetrapods – you, me, Fred the cat, cows, horses, birds and frogs – tells us that limbs are excellent adaptations for moving about on land. Also, our experience of present-day non-tetrapod vertebrates – conventionally, the fishes – tells us that these animals, adapted for life in water, have fins instead of limbs. However, somewhere in time, fishes did indeed evolve legs and start to walk on land. We might assume that limbs evolved precisely for the purpose of walking on land. This assumption, however, would be scientifically unjustified, because we can never know that it is true. After all, we weren't there to watch it happen. However, given that tetrapods plainly use their limbs for this purpose today, does not this caution seem extreme? It is not, because the fact that tetrapod limbs are adapted for walking *now* need say nothing about the reasons why limbs evolved in the first place, more than 360 million years ago.

To explain what happened in the past in terms of present-day adaptations is to tread on thin ice. It is also illogical. The example of Borges's unicorn can tell us why. To understand the evolution of the unicorn, we will first want to know where this animal stands in the tree of life, in relation to other animals, such as horses and deer. Understanding the present-day functions of the unicorn's various parts will be of no help at all in answering the question of how the unicorn is related to other animals. Arguing about whether the unicorn's horn is an adaptation for sexual display or for resolving territorial conflict, will not get us any closer to deciding whether the unicorn is more closely related to horses or deer. Nevertheless, there are plenty of real-life examples in which scientists try to do exactly that – to use present-day adaptation to explain past history.

Ultimately, a scientific explanation of the position of the unicorn in nature depends on testing hypotheses. A cladist will look at as many of the unicorn's features as are available; compare these features with equivalent (or 'homologous') features in other animals, and use these comparisons to create a cladogram illustrating the relationships between the unicorn and other animals. A cladist will be just as interested in the number of toes or teeth borne by the unicorn, as in the presence or absence of its horn.

Importantly, a cladist will try not to make any judgments of the importance of one feature or another based on its presumed function; all that should matter is that the horn is there, not whether it is used for charming damsels or impaling passing wyverns. Crucially, you should have a clear idea about the position of the organism in nature *before* speculating about the function of its various parts.

Let's say that you have discovered that unicorns use their horns to kill dragons. Using this information, you could spin a tale about the importance of the horn in unicorn evolution: unicorns evolved in dragon country, where possession of horns was an asset. Unicorns without horns would all be charred to ashes by the fire-breathing dragons. Only those unicorns with horns survived to perpetuate the species. This story sounds plausible but, like the story about the evolution of tetrapod limbs, it cannot be tested. What is more, if you use your prior (and untestable) assumption that the unicorn evolved its horn to kill dragons as a guide to the unicorn's relationships, you cannot then use this information in any subsequent test of the function of the unicorn's horn. Why? Because you have *already* assumed that you know the horn's function, even before you run the test. You have loaded the dice to tell you what you want.

The key flaw in the above story is that the central hypothesis, that the presence of the unicorn's horn played a key part in the evolution of the unicorn, is not testable. Testing hypotheses about adaptations is surprisingly hard, even when the animals concerned are living. It is impossible when you turn to fossils, the product of Deep Time. Fossils are not living creatures, but the imperfect fragments of things which we assume to have once been alive; which need not resemble any creature now living. To speculate about adaptation in extinct creatures is at best pointless, at worst recklessly misleading.

I have talked a lot about hypotheses. But what, exactly, is a hypothesis? It is a conjecture, a speculation, a hunch, framed in such a way that it can be tested. The result may or may not tell us more about the world we inhabit. The result will often, perhaps invariably, be in the form of a more refined hypothesis which can itself be tested. Because of this, science admits no absolute truth – all scientific statements are provisional. A cladogram, for example, is a hypothesis about the pattern of relationships, which can be tested in the light of new evidence. Figure 3 shows a cladogram expressing the relationship between myself and my cats, Marmite and Fred, in which Marmite and Fred are more closely related to each other than either is to me. This cladogram survived testing, in that it proved more parsimonious than a competing cladogram (shown in Figure 4) when subject to test (Figure 5). But this is not the last word. What would be the effect of adding other participants, such as dogs, horses, or even unicorns, to the cladistic analysis? It could turn out that adding new players might change the branching order of the established participants, so that, say, it becomes more parsimonious to suppose that cat-like features evolved more than once, and that Fred and I form a clade that excludes Marmite. This, too, is not the last word, and will always be subject to revision in the light of new evidence.

When a hypothesis repeatedly survives the questions we ask of it, does it become something stronger – a theory? In a sense, there is no formal distinction between a hypothesis and a theory. A theory is simply a hypothesis that experiments have not yet managed to refute, and which seems to explain many facts about the world in one inclusive framework. Theories, like hypotheses, are always provisional, and the best ones can always be supplanted. Newton's theory of gravity is a good example. It explains much about the world in which we live but breaks down in extreme situations, in which, for example, massive bodies move close to the speed of light, and become more massive as they do so. In these cases, Einstein's theory of relativity provides a more inclusive explanation of the world's workings.

But relativity would have remained a speculative hypothesis in the absence of scientists' ability to test its predictions. Relativity is now supported by a host of experimental data but these do not invalidate Newton's equations of motion so much as circumscribe their utility. The existence of relativity does not mean that apples do not fall out of trees with the sureness that they did in Newton's time – but if an apple had the mass of the Earth and moved half as fast as light, relativity would provide a neater, more complete description of events than Newtonian mechanics. Even relativity, though, is not the last word, and it is possible that a future physicist will find a better description of the world.[40] That, too, will be provisional.

The same is true not only in physics, but in all science. Hypotheses, once created, must be tested or discarded. A hypothesis without a test is like a spare groom at a wedding; he may be decorative, but his utility ends there. Scientists who study evolution are interested in the strategies creatures use to pass their genes on to the next generation. Like all scientists, they are wedded to the notion of testability, whether they are closely watching the couplings of dung-flies on a cow-pat, or monitoring the more elevated courtships of swallows, flycatchers, City traders or Amazonian tribespeople. It is helpful, even necessary, that the subjects of these experiments are alive and well. If you are studying the reproductive success of nesting birds, it is irritating if your experimental subjects don't lay many eggs in a given year; but it is not necessarily a disaster, because they may make up for it next year. But were your birds to become extinct between one year and the next, any hypothesis you might have on their reproductive habits will remain forever untestable. As such, it will be demoted from a hypothesis to a speculation, and cast out of the field of scientific enquiry.

Modern evolutionary biology is based, fundamentally, on a set of hypotheses created by Charles Darwin. Darwin's hypotheses form a body of work, the theory of evolution by a mechanism which he called 'natural selection'. The flourishing science of evolutionary biology today owes its vigour to the fact that Darwin's hypotheses were testable, and gave rise to other, more refined and more interesting hypotheses which Darwin probably could not have imagined a century and a half ago.

Darwin's idea of natural selection was informed by several different sources. As a young man, Darwin had journeyed round the world as captain's companion and naturalist on the survey ship HMS *Beagle*. The voyage of the *Beagle* exposed Darwin to the abundant biological diversity of the tropics. Another influence was a book, *The Principles of Geology*, by pioneering geologist Charles Lyell, in which Darwin learned that the Earth was immensely, perhaps immeasurably old. Lyell and his generation were the first geologists to show that the Earth was much older than implied by the Bible. In doing so, they discovered Deep Time. Given the expanse of Deep Time, Darwin reasoned, evolution would have had plenty of time to produce the diversity he observed in nature. But how? The catalyst was another book, *An Essay On the Principle of Population*, by the economic theorist Thomas Malthus.

In this book, Darwin read Malthus's view that populations tend to grow faster than the resources necessary to sustain them, and the result is deprivation. Malthus's intention was to study the causes of urban poverty, but Darwin made a bold leap, extending this principle to the whole of the natural world.

Darwin's observations told him two facts about living things. The first is their variety, both between species and within them. The second is that creatures produce far more of their varied young than can possibly survive. Most creatures – insects, fishes, plants – produce hundreds, even millions of young, only a small proportion of which reach maturity.

Darwin explained this profligacy as follows. The offspring of two parents will vary among themselves in all sorts of ways. This variation is inherited. Because there are always too many offspring to survive, the offspring most likely to succeed are those that happen to have those attributes best adapted to the prevailing environmental circumstances. The remainder, whose attributes happen to be unsuitable, will die, or at least fail to survive until reproductive age. Repeat the process again and again, and the suitable traits will gradually spread through the population at the expense of the unsuitable ones, because only those animals with the appropriate variations will survive to breed, and pass their traits on to their offspring. In this way, a superfluity of varied offspring is winnowed to leave only the ones most suited to the prevailing environment. This process of winnowing is what Darwin called 'natural selection'.

The beauty of natural selection is its simplicity. It requires just three things to work: that offspring are varied, that this variation is inherited, and that there be too many offspring to survive. This leads to what Darwin called the 'struggle for existence' – a competition for resources, be they potential mates or sources of food. The reward for success in this struggle is the possibility of reproduction passing on one's favoured traits to the next generation. This, Darwin explained, is how organisms come to be suited to their habitats: zebras living in grassland will have stripes, because only the stripiest zebras can be lost to view in the grass. Thus hidden, they will avoid the attention of prowling leopards for just long enough to raise another generation of zebras which will inherit their parents' tendency towards stripiness. Leopards living in the dappled shade of trees will have spots the colour and pattern of dappled shade, because those with different patterns, or with patterns that match dappled shade less well, might be more easily spotted by zebras, so leopards with unusual patterns go hungry and die, leaving fewer young. And if the prevailing circumstances should change, then the animals and plants would, by insensible gradations, change with them – or perish.

Darwin pondered on natural selection for twenty years before going into print. When he did, it was in response to its independent formulation by Alfred Russel Wallace, a young naturalist working in the East Indies, and who had also read Malthus. Although biologists acknowledge Wallace's contribution, the primary honour is usually conferred on Darwin, who had reached the same conclusion much earlier, and had had time to develop the idea and work out some of its implications.

Darwin's genius resides not only in the elegant simplicity of natural selection, but in the many testable hypotheses, based on natural selection, that he proposed in many books and papers. Technical reports are still being published that test hypotheses proposed by Darwin more than a hundred years ago. In 1988, for example, L. Anders Nilsson of the University of Uppsala in Sweden reported a test of a hypothesis proposed by Darwin in 1862, in a book *On the Various Contrivances by which British and Foreign Orchids are Fertilized by Insects.*[41]

Darwin wondered why certain orchids had trumpet-like flowers of extraordinary depth. The Madagascar Star orchid (*Angraecum sesquipedale*), for example, bears hollow nectar-containing spurs, or nectaries, at the base of each bloom. These nectaries grow to the remarkable length of 30 centimetres. Reasoning on the basis of natural selection, Darwin predicted that this orchid was pollinated by a species of hawkmoth with a tongue almost (but not quite) long enough to reach the puddle of nectar at the bottom of a nectary, and that the nectaries of the orchid reached their great length by an evolutionary 'race' with the moths.

Like many flowers, orchids reproduce with the help of an insect pollinator. In the case of orchids, a visiting moth will collect the pollen (containing the male sex cells) on its body, and this pollen will be rubbed off on the stigma (part of the female sexual apparatus) of the next orchid visited by the moth. Without this contact, fertilisation cannot take place. The flowers most likely to succeed are those constructed in such a way that the visiting moth is forced to rub up against the pollen-producing anthers and, on visiting the next flower, to force its pollen-covered body into vigorous contact with the stigma.

Moths know nothing of this. All they know is that flowers contain nectar, on which they feed. To the flower, the nectar is an inducement for moths to visit, but it will be wasted if the moths do not pollinate the flower in return. If, like the Madagascar Star orchid, the nectar is carried in deep tubular nectaries, the moth can reach it only by burying itself deeply within the flower and extending its tongue to its fullest. This ensures that, as it feeds, any pollen the moth happens to be carrying is rubbed off onto the stigma of the flower, and it is fertilised. In *On the Various Contrivances* Darwin wrote:

> As certain moths of Madagascar become larger through natural selection in relation to their general conditions of life, either in the larval or mature state, or as the proboscis alone was lengthened to obtain honey from the *Angraecum* and other deep tubular flowers, those individual plants of the *Angraecum* which had the longest nectaries (and the nectary varies much in length in some orchids), and which, consequently, compelled the moths to insert their prosboces up to the very base, would be best fertilized. Those plants would yield most seeds, and the seedlings would generally inherit long nectaries; and so it would be in successive generations of the plant and of the moth.

Darwin predicted that although orchid nectaries and moth tongues would evolve together, the plant would always keep one step ahead. Nectaries would always be longer than tongues, on average, to ensure that the moths got right into the flower to feed, thus depositing pollen. Flowers with nectaries shorter than the tongues of visiting moths would lose their nectar to moths that could feed without having to bury themselves inside the bloom and come into contact with the flower's sex organs. As a result, the flowers would not be pollinated as effectively, and would set fewer seed than flowers with longer nectaries.

This hypothesis was particularly bold, given that in 1862 the pollinator of the Madagascar Star orchid was unknown. It was forty years before a candidate moth was discovered. As Darwin predicted, this moth (called *Xanthopan morgani praedicta*) has a tongue that is, on average, slightly shorter than the average length of the nectary of the Madagascar Star orchid it pollinates (25 centimetres, as against 28–32 centimetres for the orchids' nectaries). This kind of relationship, in which nectaries are slightly longer than the tongues of their pollinators, forcing the insects deep into the flowers, has since been observed in other plant-pollinator pairs. Nilsson tested and confirmed the hypothesis by experimentally shortening the nectaries of deep-flowered orchids, showing how they set less seed than flowers with longer nectaries.

The accessibility of nectar in flowers with longer nectaries than average would favour the success of moths with longer tongues, leading to a kind of escalating 'arms race'. This escalation is not inevitable; were moths with long tongues to become rare for any reason, natural selection might favour flowers with shorter nectaries.

Larger moths tend to have longer tongues. The reason for a decline in large moths, and the consequent reversal of the trend towards lengthening nectaries, need not be connected with the moth–orchid relationship in any direct way. It could be the arrival of a new predator, a bird with a particular appetite for large moths. The abundance of this predator might be influenced by other factors: the bird might itself feature in the diet of some larger animal whose population also varies; or it might find itself the victim of some sudden hazard, such as an epidemic of a new virus or the destruction of its habitat.

Such factors may seem remote from moth tongues and orchid nectaries, but could influence the lengths of both. Let us imagine that a warm winter has allowed more than the usual number of small rodents to breed. This in turn encourages the fledging of buzzards, which feed on the rodents. But one species of buzzard also preys on the chicks of a species of finch, reducing the numbers of adult finches that year, compared with the year before. These finches prey on the moths that pollinate orchids, preferring the largest, juiciest specimens. A dearth of finches means that large moths are spared, and more of them survive to pollinate orchids. Because large moths are relatively abundant and have longer tongues than smaller moths, natural selection will favour orchids with longer nectaries. Next year, everything could change. Buzzards might be rarer, so that finches with a taste for large moths become more common. As a consequence, larger moths will become rarer than smaller moths, so natural selection will favour orchids with shorter nectaries, all the better to suit the dimensions of smaller moths. The ecological relationships of organisms are more than simple interactions between, say, a flower and its pollinator.

These interactions are moulded by an ever greater network of links which ultimately form the web of life – or more technically the 'ecosystem'. In the same way that the ramifications of the global economy make it hard to predict, for example, how unemployment in Texas a year from now will be influenced by a change of fiscal policy in Taiwan, the network of interactions in an ecosystem is so complicated that it is difficult to predict precisely what would happen were one particular participant within that ecosystem to become more or less abundant, or to change its habits.[42]

When evolutionary biologists try to understand adaptation – the reasons *why* organisms come to be suited to their environment and mode of life – they try to learn as much as possible about the organism's habits, behaviour, circumstances and ecological relationships. Although it may be true that the length of the tongue of a moth reflects an adaptation for supping nectar from deep orchids, one can say, with equal justification, that the deep flowers of orchids are adaptations to improve pollination by long-tongued moths and that both adaptations are influenced by the ecological circumstances in which moths and orchids live.

Hypotheses about adaptation are useful only if they can be tested. For example, L. Anders Nilsson was able to test and confirm Darwin's hypothesis about the adaptation of the tongues of moths to penetrate the nectaries of orchids. But framing reasons why certain adaptations exist is complicated by many other factors, for creatures do not live in isolation from the rest of the natural environment.

We have seen that the evolution of any structure is influenced by a multiplicity of factors, and is not a simple matter of cause and effect. To say, simply, that the hawkmoth tongue *is adapted for* penetrating the orchid nectary is to run the risk of oversimplification.

It would be like substituting a comprehensive analysis of evolution for a 'just-so' story in which adaptation is assumed to have a purpose. That the elephant has a trunk is undeniable and it is probably fair to assume that many things contributed to its evolution. However, it is almost certainly not fair to say that the elephant's trunk is adapted for squirting zoo keepers with peanuts, still less that it *evolved* for that purpose.

Misinterpretations about 'adaptive purpose' ignore the fact that natural selection is a blind and undirected consequence of the interaction between variation and the environment. Natural selection exists only in the continuous present of the natural world: it has no memory of its previous actions, no plans for the future or underlying purpose. It is not a winnowing force with an independent existence that can be personified like Death with his black cowl and scythe.

Given that adaptive purpose, as a concept, is antithetical to the essence of natural selection as blind and random, it is a surprise to learn that it finds a place in the *Origin of Species*, Darwin's most famous and influential book. In the *Origin*, Darwin used 'artificial' selection, in which breeders of domestic animals use selective breeding to promote desired traits, as a metaphor for 'natural' selection, the process that occurs in the natural world.

But artificial selection is an imperfect metaphor for natural selection because breeders quite obviously *do* have intelligible reasons for selecting some traits and not others. Unlike natural selection, breeders have memories, plans and purposes. They select for the same traits, generation after generation, to produce a discernible trend – sheep with ever-fleecier wool, beef cattle with ever-meatier carcasses, and so on.

Natural selection could hardly be more different. Unlike breeders, who reliably select for the same things time after time, the environment is always changing. Traits that are favourable in one generation may be more, or less favourable – or even detrimental – in the next. The degree to which different traits will be selected, for or against, changes from moment to moment, in response to a multitude of interacting ecological forces. I like to think of these ever-changing degrees of selection as share prices in a volatile stock exchange. Like share prices, the forces with which selection acts on traits change all the time – sometimes more, sometimes less – according to environmental pressures, and interact with one another in all kinds of ways that you could never follow, with results that you could never predict. Spotting long-term trends in evolution, as in the stock market, is difficult, especially when extrapolated from information from one single point in time – the present day. In which case, you are entitled to ask whether any trends discerned from the activities of natural selection are likely to be the product more of human imagination than reality – a question that becomes more pointed for trends that are extended over thousands of generations, and into the realm of Deep Time.

Palaeontologists face obvious problems when trying to reconstruct the lives and times of extinct animals and plants. Extinction thwarts any opportunity to perform the kinds of experiments that other evolutionary biologists take for granted, and makes any hypothesis about the biology of extinct creatures difficult to test. As a consequence, much of what we think we know about the history of life is based on untested or untestable hypotheses.

Darwin took an orchid and predicted the moth. Imagine the reverse – that you have the moth but not the orchid. If you didn't know orchids existed, how could you know why moths have such long tongues? You could guess that the tongues are used for collecting nectar but, without actually seeing moths use their tongues for this purpose, you would have no grounds for asserting this as an adaptive purpose. You could, with equal justification, propose that longer tongues improve the moths' sense of smell or that moths with longer tongues are more attractive to the opposite sex or, in fact, any other explanation that seems plausible. Adaptive purpose is something that people award to structures after the fact.

Now imagine that both long-trumpeted orchids and the moths that pollinate them are extinct and that nothing remotely like either of them lives in the modern world. Should you come across a fossil of a hawkmoth, you might be tempted to invent hypotheses to explain the adaptive purpose of its tongue but you would be unable to test these hypotheses. Your hypotheses, then, would not really be hypotheses because hypotheses imply tests. They would, instead, be what some scientists, favourably disposed to such things, call 'scenarios' – and what other scientists, less well-disposed, call 'just-so stories'.

Scenarios about adaptation contain a logical *non-sequitur*: that the adaptation of a structure for a particular purpose necessarily tells you something about how that structure evolved. Simple scenarios, say about the evolution of moth tongues, neglect the fact that orchids also evolve, and that the evolution of both moths and orchids is influenced, in turn, by the evolution of many other things in the ecosystem. Yet, were we presented with fossil moth tongues in isolation, with no fossil orchids or any other vestiges of the ecosystem in which moths and orchids lived, we would be reduced to a scenario that we could never test, still less prove. Such is the lot of palaeontologists, who must study the fragmentary fossils of unknown creatures, imperfectly comprehended through models, and divorced from their contemporary ecological contexts. This makes conventional assertions about adaptive purpose seem rather hollow.

With no ecological context, such assertions depend on present-day circumstances to explain the past. The assumption that limbs were adapted for life on land comes, ultimately, from our experience of the world around us, which tells us that tetrapods with fully developed limbs and digits live on land. In every case in which tetrapods have returned to full-time life in water, the limbs are reduced, and the digits are lost or fused into paddles. In the light of experience, then, *Acanthostega* should be an animal adapted to life on land but the presence of internal gills tells us that the animal was obliged to live under water. This conundrum can be solved only by calling into question our common association between limbs, digits and land life. Once this is done, it is no longer possible to justify any assertion about adaptation in an extinct form, based on present-day experience.

In addition to such questions about adaptive purpose, you can also never be sure, or even confident, that the environment remained constant for the millions of years necessary for natural selection to produce a discernible trend; was the onward, upward drive onto land so constant, so imperative, as to have driven the inexorable transformation of fins into limbs? Of course, we know that limbs with digits evolved, and presumably natural selection was a force in their creation but, as the example of *Acanthostega* shows, we cannot use assumptions or assertions about adaptive purpose to map out evolutionary trajectories.

As you might imagine, given our natural curiosity about the subject, discussions about human origins are particularly prone to such misconceptions. One researcher, studying a nomadic tribe, might be impressed by the sophisticated cooperation of men in hunting parties. The men might have a sophisticated system of sign language, to describe everything from the direction of the wind to the various species of prey. This researcher might then produce a scenario in which the roots of language is founded on in the cooperation of our hunting ancestors. Groups with better and more sophisticated systems of cooperation would have been favoured by natural selection as they would have brought home more meat to feed the females and offspring back in the cave. They would have survived at the expense of groups with less cohesive social interactions. In addition, successful hunting would favour a constant increase in hunting skill, driving cooperation to ever greater levels of complexity.

Such a scenario is theoretically indefensible, for two reasons. First, it makes a spurious connection between behaviour observed in the present day and an inference that such behaviour in our ancestors might have had evolutionary consequences. Second, this scenario depends on the untestable assumption that natural selection for cooperative hunting behaviour could have led, over millions of years, to language.

This same scenario might also prove indefensible, though for political reasons, to feminists, who might be prompted to devise an alternative scenario – of the evolution of intelligence and language being driven by the cooperation of sisterhoods of females gathering roots and berries.

There is no way of telling which of these scenarios is correct, or even if they are wrong. Your choice will depend solely on what you regard as plausible – which might be informed by sectarian opinion – and the skill and perceived authority of the person presenting you with this information.

A plausible yet untestable view of human origins of long standing is the 'aquatic-ape' scenario.[43] In this scheme, human beings went through a phase in their evolution in which they were aquatic. The evidence comes from superficial features of human anatomy including the distribution of subcutaneous fat, the presence of large numbers of sweat glands and the curiously hairless state of human skin when compared with our furry brethren among the mammals. It is also supported by aspects of behaviour that might otherwise be hard to explain, such as the observation that newborn babies can swim in a coordinated way immediately after birth.

Although the aquatic-ape scenario might be true, I can think of several objections. First, the evidence might be no better explained by this scenario than by any other, equally plausible scenario. For example, the distribution of fat and our relative hairlessness might be related in some general way to the regulation of temperature or metabolism. Second, it is unclear that the particular features chosen by proponents of the aquatic-ape scenario speak of an ancestry that is particularly aquatic. Many mammals love the water so much that they spend most of their lives in and around it, but they remain as furry as any mammal less aquatically inclined; in other words, there need be no association between hairlessness and aquatic habit. Presented with a skeleton and nothing else, it would be hard to tell that water voles, otters, golden retrievers, bears, beavers, platypuses, proboscis monkeys and capybaras were all capable swimmers, but that my cat Fred shies from a drop of rain as if he'd dissolve. On the other hand, whales, which are thoroughly aquatic, are also fatty and relatively hairless. However, their aquatic credentials are displayed most strongly not in these superficial features, but in their skeletons, which are profoundly modified in ways consistent with an aquatic habit: reduction of limbs, alterations to the skull and backbone and so on. Were humans to have gone through a phase in their evolution that was to any great extent aquatic, we would surely see signs of similar modification in our own skeletons. But we do not; the features chosen by proponents of the aquatic-ape scenario generally relate to soft tissues and physiology, which are not preserved as fossils.

Third, and most importantly as regards the discussion in this chapter, the evidence advanced in support of the aquatic-ape scenario concerns adaptations to circumstances in the past, when we were supposedly more aquatic than we are today. However, proponents of the scenario base their claim on the distribution of fat and body hair in humans today. The aquatic-ape scenario depends for its validity on explanations based on adaptive purpose but, if organisms are as perfectly adapted to their current circumstances as the tongue of the moth is to the nectary of the orchid it pollinates, then we should have lost all trace of aquatic adaptation and be fitted, instead, to our present-day terrestrial existence. We are not aquatic now, so why should we retain adaptations to a lifestyle we have lost?

This contradiction is a symptom of the unresolved tension at the core of all scenarios – a tension between the understanding of the way we live now, and how we got that way. It is tempting to use present-day adaptations to explain past history, but one need have nothing to do with the other. Stephen Jay Gould and Richard C. Lewontin of Harvard University explored this tension in an influential paper[44] in which they take the reader to Venice, to stand beneath the central dome of the cathedral of San Marco.

The hemispherical dome of San Marco is supported on four round arches, arranged on a square floor-plan. When arches meet each other at angles beneath a hemispherical dome, they do not fill out all the corners. Instead, they leave tapering triangular spaces called 'spandrels' (see Figure 9). Each of the four spandrels beneath the dome at San Marco bears a painting of an apostle, part of an elaborate design of religious instruction that encompasses the inside of the dome and the arches.

Gould and Lewontin consider the purpose of spandrels. Given their current use in the cathedral of San Marco, they exist to carry sacred images. But if some vandal pasted advertisements for cars and beer over the apostles, you could then claim that spandrels are supremely adapted for use as billboards. What if the spandrels were left blank? The answer is that spandrels are an unavoidable consequence of architecture, of placing a hemispherical dome on four rounded arches arranged on a square floor-plan. In themselves, spandrels *have* no purpose – they aren't *for* anything.

Figure 9. Spandrels arise as inevitable consequences of supporting a dome on arches.

Anyone who suggested that spandrels exist *for* the purpose of bearing religious paintings is putting the cart before the horse. As Gould and Lewontin note, they would be inviting the contempt in which Voltaire held Dr Pangloss, who saw a noble purpose in any situation, no matter how ridiculous the reasoning – 'our noses were made to carry spectacles, so we have spectacles'. Yet evolutionary biologists do much the same thing when they interpret any structure in terms of adaptation to current utility, while failing to acknowledge that current utility need tell us nothing about how a structure evolved; or, indeed, how the evolutionary history of a structure might itself have influenced the shape and properties of that structure. Tetrapods use their legs for walking; therefore, legs evolved *for* walking. But legs are not structures perfectly adapted for locomotion; they were the best that could be achieved in the circumstances, given that their evolution was constrained by the prior existence of fins. If tetrapods had been designed for land life from scratch, they would have axles and wheels.

The tension between past history and current utility is exposed further in an essay called *The Return of the Dancing Dinosaurs*, in which palaeontologist Robert T. Bakker looks at how our images of dinosaurs have changed with scientific fashion.[45] Although first reconstructed as giant lizards, people soon found much in them that resembled birds and mammals. As I discussed in the last chapter, dinosaurs were once seen as powerful, active animals, like mammals or birds rather than reptiles, but in the early years of this century things changed: dinosaurs slumped into the role of primeval couch-potatoes.

A dinosaur renaissance began to stir in the late 1960s and early 1970s, when John Ostrom at Yale began to doubt the dogma, and reconstruct the small carnivorous dinosaur *Deinonychus* as an agile, bird-like runner. Bakker, a student of Ostrom, took the same revisionist line with the large quadrupedal dinosaur *Triceratops*. In 1971, Bakker published a paper reinterpreting *Triceratops* not as a sluggish giant lizard, but as a fast-paced runner capable of galloping, a gait in which, in one part of the locomotory cycle, all four limbs leave the ground. This reinterpretation attracted considerable criticism. Bakker's critics (says Bakker) claimed that *Triceratops* could not have galloped, because it was too big.

The largest galloping animal extant is the adult bull white rhinoceros, which has a mass of about three tons. Elephants, which may have a mass of up to 7.5 tons, do not gallop. If they tried, they would break their own legs. A large specimen of *Triceratops* could have had a mass of ten tons, rather more than the largest elephant that ever existed. By the argument of Bakker's critics, *Triceratops* would have been a slow plodder, unable to gallop.

This argument fails because it judges the past by the standards of the present. *Triceratops* lived 70 million years ago, in the Cretaceous period. We know it only as a fossil, imperfectly comprehended through modern models and divorced from its contemporary ecological context. We cannot assume that the world inhabited by *Triceratops* was like our own, just as we cannot assume that adaptation to current utility says anything about the forces that have shaped forms during their evolution. By the same token, we cannot simply conceive of *Triceratops* as a magnified elephant or rhinoceros; it was a qualitatively different sort of animal. It is therefore illogical to constrain the range of activities of this extinct form on the basis of the capabilities of a set of completely unrelated animals living 70 million years later.

Bakker makes a good case that *Triceratops* was a galloper, but he is forced to rely on a series of disparate modern models, none of which are very like *Triceratops*. First, he reconstructs the *Triceratops* elbow joint as tucked in under the body, much as is found in a large galloping animal such as the rhinoceros, rather than splayed out at the sides, as in crocodiles and other reptiles, which do not gallop. The *Triceratops* elbow looks (and would have worked) more like a rhinoceros elbow than a crocodile elbow.

In modern gallopers, the spine is held horizontally, and the front legs are more or less the same length as the back legs. If the front legs were much shorter than the back legs, the animal would tip itself up on its nose. *Triceratops* presents a problem, because its front legs are substantially shorter than its back legs. This problem is solved by thinking of the *Triceratops* shoulder girdle as behaving like an upwards extension of the front limb, an arrangement that is seen in lizards such as the chameleon. The length of the front limb plus the length of the strap-like shoulder blade almost equals the ground-to-hip length of the hind limb. We see *Triceratops*, then, as a chimera made of a rhinoceros (which is, incidentally, only very distantly related to *Triceratops*) and a chameleon (which is a closer relative, but is much smaller and moves more like an ordinary lizard than a galloping animal).

Galloping animals need a great deal of hind-limb thrust. This can be estimated from the size of the attachments for the muscles, preserved as scars on the bones of the hind limbs. *Triceratops* has relatively larger muscle attachments than found in rhinoceroses or elephants. This suggests that *Triceratops* would have been a more powerful runner, judged weight for weight, than an elephant or a rhino. Interestingly, though, the arrangement of the musculature of the *Triceratops* hind limb looks rather more like that of a large bird than a rhino. This is not surprising, as dinosaurs are more closely related to birds than to mammals such as rhinoceroses and elephants, but it presents Bakker with a frustrating conundrum. A better model than a rhinoceros for demonstrating the galloping gait of *Triceratops* would be a giant, four-footed bird. But every bird we know about is bipedal, and galloping is a four-footed exercise.

Bakker presents as convincing an argument as he can, based on modern models, that *Triceratops* could gallop, but the fact remains that this extinct creature is unlike anything alive today. Some of its features such as its distinctive arrangement of horns, and the bony neck-frill, a massive rearward extension of the skull defy interpretation by any model. Without a modern model, it is hard to make sense of such idiosyncratic features.

Even so, it is still possible to get a basic, broad understanding of what *Triceratops* was like, by working out its evolutionary relationships, without any need to make conjectures about how it was adapted to its existence. As we were reminded at the end of the last chapter: before speculating about the function of a unicorn's horn, you should take time to understand the evolutionary relationships of the unicorn with other animals. Once you have done this you will have a basis for understanding how the unicorn came to be the way it is.

Triceratops has a backbone made of vertebrae, so it is a vertebrate, just like you and me, pteraspid fishes and Fred the cat. But unlike pteraspid fishes which have at most a single pair of paired fins, *Triceratops* has four limbs, the bones of which correspond with those in lobe-fins such as *Eusthenopteron*. As a vertebrate, *Triceratops* shares a closer common ancestry with lobe-fins such as *Eusthenopteron* than it does with a pteraspid. The limbs of *Triceratops* carry digits, so it is a tetrapod and shares a closer common ancestry with the tetrapod *Acanthostega* than either does with *Eusthenopteron*, which does not have digits.

The bones in the skull of *Triceratops* were arranged in a particular way suggestive of a shared common ancestry with birds, crocodiles and lizards such as the chameleon. In particular, the cheek region behind the eye sockets in these animals tends to be perforated with two large, distinctive holes. In mammals such as the white rhino and Fred the cat, and in the extinct reptilian relatives of mammals, the skulls are arranged differently, with only one hole in the cheek region. Therefore we can say that *Triceratops* and the chameleon share an ancestry that excludes mammals. Given that chameleons and dinosaurs other than *Triceratops* are known to have laid eggs, it is a reasonable assumption that *Triceratops* also laid eggs. This argument is easily summarised in the cladogram in Figure 10.

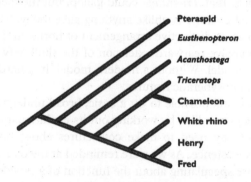

Pteraspid

Eusthenopteron

Acanthostega

Triceratops

Chameleon

White rhino

Henry

Fred

Figure 10. A cladogram showing the position of *Triceratops* in the pattern of life.

Being able to place *Triceratops* in a cladogram allows us to compare this extinct form with animals about which we know more. In so doing, we start to understand *Triceratops* a little better – importantly, without making any prior assumptions about how or why *Triceratops* evolved. The cladogram shows that *Triceratops* is only a distant relative of the rhinoceros, for all that *Triceratops* may have galloped like a rhino. Each has a different evolutionary heritage, governed by different circumstances and contexts. No rhinoceros lays eggs: the image of *Triceratops* as a ten-ton egg-laying rhinoceros, like nothing on Earth today, underlines the point that *Triceratops* was a creature of its time, living in an ecological milieu that we can understand at best only extremely imperfectly. We can never know the forces and circumstances that shaped the *Triceratops*' horns, its bony neck-frill, or even its galloping gait. To speculate that the horns and neck-frill of *Triceratops* were adapted for doing battle with *Tyrannosaurus rex* is pointless, because we can never prove it.

Even though we know little about the ecology of the world in which *Triceratops* lived, we can assume that it was as complex as the web of life in the world today. Given that few creatures fossilise, we remain largely in ignorance of the creatures that shared *Triceratops*' world, and which may have had a profound influence on the adaptations of the animal itself. In the rest of this chapter I shall illustrate this point with the converse – that there are many creatures alive today that influence the lives of their fellows, and that many of these influential creatures have left little or nothing in the way of fossils.

Take, as an example, the roundworms, or nematodes. There are at least 10,000 species of roundworm, many of them parasitic. As many as fifty different species of parasitic roundworm are known to have their existence inside living human beings. Most of the time we are hardly aware of the fact, though a few species cause serious disease. *Trichinella spiralis*, for example, is responsible for the agonising consequences of eating infested, undercooked pork. The hookworm (*Necator americanus*), scourge of the American Deep South, is a roundworm, as are the guinea-worm (*Dracunculus medinensis*) and the filarial worm (*Wuchereria bancrofti*), responsible for the disfiguring condition called elephantiasis. Large sections of society in many countries are afflicted with one or other of these parasites, with serious consequences for individual health and economic prosperity. It can hardly be denied that parasites have a huge effect on our lives. In general, though, we live our lives ignorant of the sheer abundance of roundworms, both parasitic and free-living. Millions teem in every spadeful of garden soil, and they can be found in the drinking water of even the cleanest cities. If all the matter in the Universe were made transparent, all except for the roundworms, we would still be able to discern our planet's face, the landscape, the trees and plants, and the animals, from the ghostly fog of their burden of roundworms. Towns and cities would be recognisable by moving knots of roundworms in human shape, each one the catalogue of infestation in each one of us.[46]

You would hardly imagine the present-day ubiquity of roundworms from their fossil record, which is meagre. Fossilised roundworms have been found preserved in amber, a remarkable but rare medium, the petrified resin of coniferous trees; others have been found in fossilised faeces. As far as can be told, all fossil roundworms found to date are indistinguishable from modern forms; all they tell us is that roundworms have existed for millions of years.

Similar stories can be told for other modern groups of organisms, ranging from tapeworms to viruses. All such organisms are ubiquitous in the modern world; all influence the lives of other creatures in important ways; all have exiguous or absent fossil records. Many of them are parasites or agents of disease.

The world of *Triceratops* was different in detail from the world today but we have no reason to suppose that the ecology of the Cretaceous Period was any less complex. The structural sophistication of the Cretaceous forms we know about, such as dinosaurs, is sufficient assurance of that. We can therefore assume that the Cretaceous world teemed with ubiquitous but rarely fossilised creatures, which influenced the lives, adaptations and ultimately the evolution of those creatures, such as *Triceratops*, which have been preserved as fossils.

If this view is correct, we may suspect that many of the features we see in fossil creatures were adaptations to forces we cannot possibly guess. If tetrapods evolved legs with digits for walking on land, they could equally well have evolved them to escape some water-borne infestation.

Parasitism and disease could have been responsible not just for unseen inconvenience; they could, in fact, have been the stimuli for the evolution of the diversity of life on Earth. *Triceratops* could have had its burden of worms, but without infectious agents there might have been no *Triceratops* at all, and very little else in the way of complex life forms. This conclusion seems strange but it stems from an argument about how the evolution of the immune system, and ultimately of sexual reproduction, might have been driven by the presence of disease.

The argument runs as follows. Evolution is driven by natural selection, but this is not very effective without genetic variation. Variation is maintained by sex, in which two individuals meet to exchange or combine genetic information. Sex is a mechanism for shuffling genes in each generation. Although sex explains how variation is maintained, it does not tell us *why*. The origin of sex remains an important problem for evolutionary biologists.

One answer is that variation is an effective hedge against disease. When a new disease strikes a genetically varied population, some of the individuals in the population are likely to have a natural immunity to the disease, and will therefore survive to reproduce. Asexual populations, or populations that are less genetically varied, will be more prone to extinction. Although it is easier to reproduce asexually, or 'clonally' – all asexual creatures need to do is produce copies of themselves, rather than go to all the bother of exchanging genetic material with another individual – the effort of sex may be the price for keeping variation as high as possible.

The above argument sounds like an adaptive scenario, and it is – but it is supported by a wealth of evidence. Some of the most intriguing evidence concerns the way organisms choose their mates. Mice, for example, tend to select mates that are as different as possible, genetically, from themselves. They gauge this difference from pheromones, secreted in the urine, that are derived from the genetically varied system of proteins used by the immune system for telling the difference between cells of the body and cells of an interloper, such as an infectious agent. In this way, mate choice is specifically and directly influenced by genetic variation maintained as a hedge against disease.[47]

It could be said that sexual reproduction evolved as a way of staying one step ahead of disease. Because of sex, natural selection has more variation to work on than would be the case were creatures routinely to breed asexually, by budding or simply dividing in half, like an amoeba. The result is that the pace of evolution is quickened in sexually reproducing organisms.

The earliest fossils are of bacteria, found in the 3.6-billion-year-old Apex Chert of Australia, first described by J. William Schopf.[48] Bacteria are usually (although not always) asexual. Between 3.6 and about 2 billion years ago, evolution was very slow. Schopf has suggested that some of the organisms preserved as fossils more than 3 billion years old are virtually identical with organisms, called cyanobacteria, living today.[49] Nobody knows when sex first evolved but the appearance in the fossil record of organisms thought to have been capable of sexual reproduction led to a rapid increase in the pace of evolution. A burst of evolution around a billion years ago led to a wide variety of microbial life. Another burst, around 600 million years ago, produced, for the first time, animal forms large enough to be seen with the naked eye, and which left fossilised remains.

It is therefore possible, indeed likely, that much of the evolution of life on Earth has been driven by an evolutionary response to the threat of parasitism and disease. That diseases are caused by organisms that rarely, if ever, have a fossil record, underlines the point made above: that you can never make confident assertions about how adaptations in organisms promoted this or that course of evolution, observed after the fact, from the limited range of fossils available. A comprehensive knowledge of the ecological milieu of organisms is necessary. In the case of fossils, this is hardly ever available, as some of the most important determinants of evolution, such as the agents of disease, are forever hidden from view.

Negotium perambulans is a short story by E. F. Benson in which residents of a house built on the site of a desecrated church were relentlessly pursued by a black, blood-sucking apparition, bent on exacting divine vengeance, that could be confined to the shadows only as long as there was oil to light the lamps. The key to the horror lay in an obscure antique panel. The caption read *negotium perambulans in tenebris*, quoting Psalm 91, translated in the King James

Version as 'the pestilence that walketh in darkness'. The picture showed a hideous man-sized leech.

This obscure gothic tale could be a metaphor for the role of unseen forces in shaping the history of life. For billions of years, living things have been engaged in a staged retreat from the pestilences that walk in a darkness so deep as to deny even the possibility of fossilisation. The history of life – the evolution of sex to maintain genetic variation, and the subsequent flowering of diversity that this engendered – could be read as a constant struggle to keep one step ahead of the unseen agents of disease. The fossil record would give us a better picture of this history of life were all animals preserved along with their pestilences. The fact that they are not should be warning enough for us not to read in too much about the lives and habits of extinct creatures from the scraps that they may, occasionally, chance to leave behind.

4 Darwin and His Precursors

> In each of these texts we find Kafka's idiosyncrasy to a greater or lesser degree, but if Kafka had never written a line, we would not perceive this quality; in other words, it would not exist . . . The fact is that every writer *creates* his own precursors.
>
> Jorge Luis Borges, *Kafka and His Precursors*

Why did the dinosaurs become extinct? The apparently sudden disappearance of the dinosaurs from the fossil record at the end of the Cretaceous Period, 65 million years ago, has excited comment for decades. Palaeontologist Mike Benton has compiled a bibliography of hypotheses to explain dinosaur extinction, culled from the scientific literature.50 The catalogue is both amusing and amazing. Dinosaurs died out because the climate became too hot, too cold, too wet, too dry, or a combination of the above; because their eggs were eaten by mammals; because their eggshells became too thin so that they broke before hatching; because their eggshells became too thick – a consequence of great size – so that the young could not hatch at all; because of indigestion caused by eating flowering plants, which first appeared in the Cretaceous Period; because of hay-fever, brought on by exposure to pollen from these new flowering plants; because they contracted a deadly virus; because of radiation from a supernova elsewhere in the galaxy; because of the climatic after-effects of an episode of violent, global volcanic activity; because the Earth was struck by an asteroid; because, having been dominant life-forms for 150 million years, the dinosaurs just got bored.

The current consensus is that the extinction of the dinosaurs was related to the devastation caused by the impact of an asteroid at the very end of the Cretaceous Period. Much evidence supports the hypothesis that an asteroid landed at the required time and caused environmental upset of the required magnitude to have done for the dinosaurs.[51] However, establishing whether the impact really *did* extinguish the dinosaurs – making a link between cause and effect – is impossible, given what we now know about Deep Time.

The sheer industry with which people searched, and continue to search, for causes, is itself intriguing. If dinosaurs became extinct, people argued, there must have been a reason *why*. But to link the extinction of the dinosaurs with any particular cause – to ask questions such as 'how' and 'why' – is to demand of Deep Time more than it can answer. Dinosaurs are fossils, and, like all fossils, they are isolated tableaux illuminating the measureless corridor of Deep Time. As we saw in Chapter 1, no fossil is buried with its birth certificate. That, and the scarcity of fossils, means that it is effectively impossible to link fossils into chains of cause and effect in any valid way, whether we are talking about the extinction of the dinosaurs or chains of ancestry and descent. Everything we think we know about the causal relations of events in Deep Time has been invented by us, after the fact. Although the extinction of the dinosaurs is undeniable, the cause will forever remain elusive. It could have been an asteroid, a series of volcanic eruptions, or a multitude of disparate causes each inconsequential on its own.

The most interesting point to emerge from Benton's analysis was that until the 1950s the subject of dinosaur extinction attracted comparatively little interest. The view used to be that the extinction of the dinosaurs was part of the natural order of things; there was no need to search for its particular cause. The dinosaurs died out as day follows night; they appeared on the stage of the world, played their parts, and left, to be supplanted by the mammals. The fate of the dinosaurs, as for all animals, was seen as foredoomed, inevitable, as if the courses of evolution were laid down from the beginning of the world and animals evolved by following these pre-set tracks. The popular view of evolution is still of a chain of progressive improvement, whose course is preordained, so that each fossil discovery can be seen as a 'missing link' just waiting to be discovered, a piece in a grand jigsaw puzzle whose solution already exists.

Even today, when films such as *Jurassic Park* portray dinosaurs as active and intelligent, there is still a view ingrained in popular culture that the dinosaurs died simply because their time had come. After 150 million years as the Lords of the Mesozoic Era, they were obsolete, destined for the scrapyard of history, outclassed by quicker, brainier mammals. The mammals would have owed their superiority to their later appearance in the fossil record – a necessary correlate of progressive evolution: for if mammals had *not* been superior to dinosaurs, they would never have been able to replace them. It therefore follows that, by comparison with mammals, dinosaurs would have been dim, shuffling and inadequate. This idea persists today in the use of 'dinosaur' as a pejorative term meaning lumbering and obsolescent, outmoded by the arrival of newer, leaner, more dynamic versions, whether of computers, cars or corporations.

This view of evolution is seen everywhere in advertisements.[52] A recent TV commercial for a car showed a shiny new example purring through the landscape. The voice-over extolled the virtues of the new model over the competition and earlier versions of its own marque. The conclusion – the catch-line meant to stay in the mind of the viewer – was 'It's Evolved'.[53]

As noted above, it is only natural that we should be interested in our own origins. The narrative account of our own evolution tends to emphasise the part played by our own ancestors, editing out any side-branches that led to the evolution of other creatures. This gives a false linearity to the picture, an impression of inevitable progression. From our vantage point in the present, we arrange fossils in an order that reflects gradual acquisition of what we see in ourselves. We do not seek the truth – we *create* it after the fact, to suit our own prejudices.

An illuminating literary parallel to this retrospective creation of history appears in Borges's essay *Kafka and His Precursors* which investigates the task facing the critic seeking to discover the literary influences on any given author, in this case Franz Kafka. Borges starts by discussing a disparate selection of works in which one might recognise Kafka's voice. They include Zeno's paradox against movement, which Borges claims is cast in a form similar to that of Kafka's story *The Castle*; an ancient Chinese discourse on the inaccessibility of the unicorn (the source of the epigram that heads

Chapter 2); the writings of the philosopher Søren Kierkegaard, the poet Robert Browning, the gothic fantasist Lord Dunsany, and others.

In all these heterogeneous pieces, says Borges, we can recognise the form, style or tone of Kafka but, apart from that, none of these pieces resembles any of the others. If Kafka had become a plumber instead of a writer, we would have had no cause to discuss Zeno, Browning or Lord Dunsany in the same paragraph. The connections between all these works are made, after the fact, in the mind of the reader. Furthermore, the way we read any of these works is profoundly influenced by our having *already read* Kafka, so we are attuned to any Kafkaesque phrases in a way inaccessible to the authors concerned. 'The poem *Fears and Scruples* by Browning foretells Kafka's work', says Borges, 'but our reading of Kafka perceptibly sharpens and deflects our reading of the poem. Browning did not read it as we do now'.

The impact of Darwin's views on modern thought has been so profound that it is extremely hard for us, today, to imagine how people thought about the history of life before the publication of the *Origin of Species* in 1859.[54] We think of every aspect of our lives in terms of phrases such as 'the survival of the fittest', 'the struggle for existence' or even 'it's evolved'. Darwin's position of eminence is, in fact, so great that it transcends the immediate particularities of commemoration. Scientists such as Faraday and Newton have appeared on British banknotes: not so Darwin. The achievements of Newton and Faraday can be easily grasped; the difference that they made is palpable.

Darwin's victory, on the other hand, has been so complete that we cannot imagine a time before Darwin, so commemoration seems irrelevant. Darwin is just *there*, and will be so for ever, without our having to erect statues in his honour. No splendid thoroughfare or elegant London square need carry Darwin's name: his memory is assured. So much so, that until recently, Darwin's home, Down House in Kent, languished in slow decay for lack of funds.[55]

Evolution by copywriter – ad-man's evolution – is entirely at variance with the random, undirected process of natural selection as described by Darwin. As we saw in Chapter 3, natural selection, unlike artificial selection, has no memory or purpose. There is nothing inevitable or progressive about it. Evolution carries the fates of organisms where it may, according to the prevailing circumstances, without recourse to any pre-existing plan. Any trends we see in evolution are ours to make; as Borges says, every writer *creates* his own precursors. Or, in the portentous words of Larry Gonick's cartoon fish, heaving itself ashore, '*my* descendants will write the *book*'.

There are, then, not one but two views of evolution current in the modern world. One is the Darwinian view of evolution, driven by contingent adaptation and blind natural selection; the other is the view of evolution as directed and progressive. These views are mutually antithetical. Neither can tell us much about the history of life over Deep Time. On the one hand, to use the present-day adaptations of creatures as explanations for their history is both illogical and scientifically invalid; it simply does not follow that the uses to which a unicorn turns its horn should tell us anything about the evolution of that same structure, or the position of the unicorn in the history of life. On the other hand, to take the blind and undirected process of natural selection and turn it into an instrument of destiny is simply perverse.

The failure of both views of evolution rests, once again, on the failure to understand that Deep Time cannot sustain scenarios based on narrative. I return, once again, to the thought experiment that is central to my argument: next time you see a fossil, ask yourself whether it could have belonged to your direct ancestor. Of course, it *could* be your ancestor, but you will never be able to know this for certain. To hypothesise that it might be your ancestor, then, is futile, because your hypothesis would be untestable. So, to take a line of fossils and claim that they represent a lineage, is not a scientific hypothesis that can be tested, but an assertion that carries the same validity as a bedtime story – amusing, perhaps even instructive, but not scientific.

In this chapter I shall look at how these two antagonistic yet strangely intertwined views of evolution came to be. Ultimately, they represent a confusion between pattern and process, between the shape of the tree of life and the forces that created it: an expression of the tension that arises from our failure to understand that Deep Time is a qualitatively different medium from ordinary, everyday time. By the end of the chapter you should have a good idea what cladists are up against when they seek to create a new view of the history of life that is sensitive to the challenging scale and unique properties of Deep Time. To understand this tension, we must try to see the natural world through the eyes of someone schooled in the classical tradition of natural history, who had never heard of Darwin or evolution.

Should you find yourself at Arlanda International Airport in Stockholm, you might consider stopping for refreshment at the Bar Linné, named after the eighteenth-century Swedish naturalist Carl von Linné, better known as Linnaeus.[56] It was Linnaeus who organised the discipline of taxonomy – the formal system of latinised nomenclature by which nature is organised into species, genera and so on – into something like its modern form. The tenth edition of Linnaeus's *Systema Naturae*, published in 1753, is regarded by convention as the Year Zero of taxonomy. Formal names of species created before that date are not recognised. But names created after 1753 are subject to the rules and regulations of taxonomy (and these are many and detailed) that remain in force to the present day. We owe the name for our own species – *Homo sapiens* – to Linnaeus.

To Linnaeus, all species were viewed as having been created separately, as set down in the Bible. No new species could have evolved afterwards – the very concept was meaningless – and the number of species was finite and unchanging.

Linnaeus and his colleagues were trying to work out how best to arrange the species they found in nature into a system that would best illuminate the underlying plan of the Creator. An orderly system of nature would have a place for everything, and everything would have its place. Ideally, the organisation of nature would result in a scheme similar to the Periodic Table – the familiar chart as seen on the wall of every school science laboratory. In the Periodic Table, the elements are arranged to reflect a degree of similarity as regards their properties; an arrangement that in turn reflects their under-lying structure. One group of elements, the alkali metals, for example, contains (in order of increasing atomic weight), lithium, sodium, potassium, rubidium and caesium. These are all shiny, silvery metals, but their extreme chemical reactivity means that they are rarely seen in their native state, uncombined with other elements. When they do combine, they do so with extreme force, to form partnerships or compounds called 'salts' which are crystalline materials; the most familiar is common or rock salt, sodium chloride. The high reactivity of alkali metals is explained by atomic structure. The atoms of any alkali metal have a single, lone electron in the outermost of one or more electron 'shells', and this electron is readily shed or 'donated' to atoms of elements with vacancies in their outermost electron shells.

The order in which the elements are arranged in the Periodic Table reflects their affinities. When elements with similar properties are grouped together, regularities in their underlying structure, such as the numbers of electrons available to participate in chemical reactions, are revealed. Such regularities have great predictive power. When the Periodic Table was first invented by the Russian chemist Mendeleev, element 43 was yet to be discovered. Even though not a gramme of this element was known, many things about its behaviour could be predicted from the Periodic Table. In time, this element – technetium – was discovered, and found to behave as predicted. Today, when nuclear chemists synthesise atoms of novel chemical elements in nuclear reactors, the chemical properties of these elements can be confidently predicted, thanks to the Periodic Table.

The order of the Periodic Table reflects the affinities of elements to a degree so refined that it allows the prediction of the existence of hitherto unknown elements. Importantly, the affinities between the chemical elements in the Periodic Table do not reflect any genealogical connection. Just because alkali metals are arranged in a certain sequence, starting with lithium and progressing through sodium, potassium, rubidium and ending with caesium, this arrangement does not imply a course of evolution from lithium to caesium; it reflects no more than a graded series of properties.

There is an important difference between the elements as listed in the Periodic Table and instances of these elements in nature. The element carbon, with an atomic number of 6 and a place in the

Periodic Table above the element silicon, with boron (element 5) to one side and nitrogen (number 7) to the other, is only a cipher. It has no existence in the real world. It is an *idea* that stands for all the many instances of carbon we see around us, whether on its own, as in diamonds, graphite, coal or soot, or in combination with other elements, as in petroleum, plastic, chalk, lumps of sugar or the bodies of organic beings. All these are natural instances of the element carbon, but the element itself stands above and beyond all these as an ideal state – an archetype.

Linnaeus and his contemporaries saw the organisation of nature in much the same way as we see the Periodic Table of the chemical elements. It is not often appreciated that Linnaeus incorporated minerals into his system of nature, as well as animals and plants. This seems very strange to us today, given that we see everything through the eyes of evolution. After all, animals and plants evolve, but minerals do not. This distinction would have meant nothing to Linnaeus, who lived and worked more than a hundred years before the *Origin of Species*. To Linnaeus, species of animals and plants were as fixed, as immutable, as any species of mineral. Individual animals such as Fred the cat were instances of a general category, 'cat', an archetype that existed only in the plan of nature or the mind of God.

In this, the classical view of nature, organisms were allowed to vary, even to the extent of moving from one archetype to another, but the archetypes themselves remained fixed and separate from the instances – that is, the real-life examples of organisms – they represented. To pursue the analogy of the Periodic Table even further, uranium changes naturally into lead through the process of radioactive decay. But this transformation applies only to those instances of uranium – those particular uranium atoms – involved in that particular episode of radioactive decay. It does not apply to the *idea* of the element. If a few particular atoms of uranium happen to decay into a few particular atoms of lead, the archetypal concepts of 'uranium' and 'lead' remain unchanged. In the classical concept of nature, by analogy, the *idea* of cats persisted and remained separate and immutable, no matter what happened to any particular cat or group of cats. In other words, there were two distinct notions of the species: the instances seen in the normal course of everyday events, and the unseen 'archetypes' to which these instances conformed.

It is sometimes thought that species in the classical world were fixed. This is not strictly true; immutability was a property only of the archetype, not of its instances. But if organisms did vary in any classical scheme, they did so according to prescribed tracks dictated by affinity. Fishes, if they changed at all, would change into amphibians, amphibians to reptiles, and so on. This is precisely analogous to patterns of radioactive decay as observed in the chemical elements. Radioactive uranium always changes into lead, invariably through the same sequence of intermediates (all of which are known chemical elements), at precisely predictable rates, according to known physical laws. The paths of radioactive decay are predictable, unwavering and immutable. Uranium does not decay into carbon one day, xenon the next, and lemon meringue pie the day after, according to the prevailing circumstances. Classical natural science demanded the same degree of regularity from nature as we now do from the chemical elements or other physical systems. There seemed no good reason why this demand might have been unreasonable.

To complete the analogy, Linnaeus and his contemporaries drew up versions of the natural world that expressed the properties I have outlined. Such a scheme would have been a Periodic Table of archetypes of the natural world. In general, we would call such a plan a Great Chain of Being. The archetypes were placed next to one another according to affinity and usually arranged in order of increased perceived organisational complexity. Simple animals such as polyps might be placed at the bottom, followed by worms, insects, molluscs and vertebrates, the vertebrates being subdivided into fishes, amphibians and reptiles, mammals and birds. This arrangement, in order of complexity, no more implied evolutionary change than ordering a group of elements in order of atomic weight.

The classical view of nature had stood essentially unchanged since the days of Aristotle and Plato. The first cracks in this philosophy appeared in the eighteenth and nineteenth centuries when people began to realise that fossils were not freaks of nature but the remains of real creatures that had vanished from the Earth, perhaps in Noah's Flood or a similar catastrophe. This led the German-born French savant Georges Cuvier, excavating the bones of strange, hitherto unknown mammals from the rocks of the Paris Basin, to the idea that species were not permanent, but could become extinct.

The idea of extinction was revolutionary, given the conventional wisdom that species were fixed and unchanging for all time. Many people clung to classical ideals for many decades. Even Darwin's contemporary and geological mentor, Charles Lyell, supposed that those species that had vanished – pterodactyls, say – might still exist in some unexplored region or, if they were completely extinct, might one day re-evolve in precisely the same forms, as if following a preordained, archetypical plan. This makes perfect sense in the classical scheme. It is like saying that if we ever ran out of bicarbonate of soda, we could reasonably suppose that it existed elsewhere in the Universe, but even if we did not find any, we could still make some by the appropriate chemical reaction.

Such is the view of the world that Darwin – wittingly or unwittingly – annihilated. As we have seen in Chapter 3, Darwin's genius lay in the simplicity and elegance of natural selection as a mechanism to explain evolutionary change. But this would not have been possible without a more fundamental advance. Before natural selection can be made credible as a force for change, it is first necessary to abolish the archetype.

In the classical world, instances of a species varied, to a greater or lesser extent, from the archetype. This variation was seen as an inconsequential side-effect of the imperfection of the real world, when compared with the Platonic perfection of the archetype. Darwin moved variation from being an incidental detail to the very essence of his theory. It was variation that provided the raw material for natural selection; variation was not a fact of the imperfect world at which sages would nod their heads with resignation, but a vital part of life without which evolution could not happen.

Darwin saw evolution as a process of slow but seamless change. Through the action of natural selection, populations of creatures would slowly change in form and habits, to keep in step with the changing environment. Over millions of years, one species would slowly transmute into another. In the classical world, species could also change from one to another, but they would be constrained to change between defined archetypes along predetermined courses. Darwin saw no need for archetypes. It was far simpler to imagine species changing simply as a consequence of natural selection – of variation interacting with the environment. As a result of this, species could change in unpredictable ways, governed by the interaction of variation with the ever-shifting, endlessly complex ecosystem.

To a natural historian in the Linnaean tradition, Darwin's idea must have seemed quite bizarre. To pursue the Periodic Table analogy, it was like saying that uranium need not always decay into lead, following the prescription of physics. It could accrete electrons and protons or shed them, according to what best suited it at the time, and evolve into absolutely anything. It was like saying that carbon might evolve into nitrogen by way of an indefinite number of new elements somewhere between the two, with fractional atomic numbers between six and seven. In promoting variation to a central feature of his theory, Darwin not only abolished the archetype but utterly smashed the natural order of things, in which every creature, including mankind, had its preordained station.

Nobody would ever accuse *The Origin of Species* of being a riveting read. I have read parts of it several times, but it has to be said that the book taken as a whole is turgid and much too long. It is full of sentences that contrive to be cramped as well as lengthy. Over-stuffed with examples, asides and relative clauses, sentences turn into paragraphs, so that by the time you reach the end, you wonder what was going on at the beginning. I used to think that mid-Victorian readers were used to books like that, and that people today have shorter attention spans, but I have changed my view. To the modern reader, the *Origin* is, plainly written, but reads like a long catalogue of the doings of flowers, moths, pigeon-breeders and so on, making a point that we already know – that evolution happens, and is driven by natural selection. But how do we know this? Because of the *Origin*, of course. Darwin's ideas are so much a part of our world-view that we take them for granted, so much so that when we actually read the *Origin*, it does not seem fresh and iconoclastic, but dreary and derivative. As Borges wrote about Kafka, we cannot read Darwin through his own eyes or those of his contemporary readers. It is therefore difficult to understand just how shocking the *Origin* must have been at the time.

Imagine that you are one of those eager readers. As you turn page after page, you will see a brilliant author at work. Slowly, carefully, and using plain, ordinary language, he strips away everything you hold dear, exposing the foundations on which your beliefs rest and then, without mercy or pause, destroying those to leave a squalid, mindless 'struggle for life' in which your view of orderly, divine creation does not exist, and never has existed.

Darwin replaced an orderly, preordained pattern of nature with a simple, guiding process that explained how that pattern might have been a consequence of randomness and makeshift, acting now as it has done since the beginning of the world.

Darwin required variation as the stuff of natural selection, but was unable to explain how variation was maintained. With no mechanism to create variation, natural selection would gradually weed it out to leave bland homogeneity, and evolution would cease. The answer to Darwin's problem was the discovery, long after his death, of particles of inheritance called genes.[57] As explained in the last chapter, variation is maintained, in part, by sex, in which genes are shuffled in each generation to produce novel combinations. As the nineteenth century turned into the twentieth, studies on evolution and the new science of experimental genetics developed along separate lines and they came together only in the 1930s, largely thanks to the work of the naturalist-turned-geneticist Theodosius Dobzhansky.

Dobzhansky's book, *Genetics and the Origin of Species*, published in 1937, represents perhaps the first fully worked-out discussion of Darwinian natural selection in terms of the behaviour of genes in populations. It marks what historians of science call the 'Modern Synthesis' of genetics and evolution. Modern textbook views of evolutionary biology are referred to as 'neo-Darwinian' and are substantially based on the views of Dobzhansky and his contemporaries, working around the time of World War II.[58]

One of Dobzhansky's contemporaries, and a fellow architect of the

Modern Synthesis, was the biologist Ernst Mayr, who in 1942 tackled the difficult problem of the meaning of the term 'species'. To classical thinkers, the concept of the species was not at all problematic. Each species was represented by an archetype, comprising a collection of more or less varied instances. Whether dalmatian or dobermann, retriever or rottweiler, a dog is a dog because it is an instance of an archetype. A dog will always be different from, say, a wolf or a jackal, which will have their own, distinct archetypes.

But if variation is the substrate of evolution, and if the individuals and populations that constitute species can vary so much that they can evolve by infinitesimal stages into other species, it becomes very difficult to define a species in an unequivocal way, based solely on the appearance of the individuals within it. If species do not change, a frog is a frog no matter how strange an individual frog might look. But if species can change, it is possible that some frogs represent evolutionary stages between frogs and other things, such as toads or newts. Defining what is meant by 'frog' in that sense becomes difficult and sometimes arbitrary.

Mayr sought to solve this problem by setting out the 'biological species concept', which says that 'species are groups of actually or potentially interbreeding natural populations, which are reproductively isolated from other groups'.[59]

The key concept here is reproductive isolation. By this, Mayr means that individuals in a species will tend to mate with other members of their own species rather than with members of a different species. This builds on an important Darwinian scenario about the process of speciation – how new species evolve. You might imagine a population of a species that becomes broken in two by a geographic barrier, such as a river or a range of mountains; or perhaps some members of the species have become stranded by accident in some spot remote from the home range. Members of the sundered populations are no longer able to mate freely with any other member of the same but separated species; geographical isolation implies reproductive isolation. Thus separated, each population will become prey to its peculiar environmental circumstances, and will evolve in its own way. Eventually, two species will form, where there was one before. Should two individuals of these two 'daughter' species happen to meet in the future, they will find that they are unable to mate or, if they are, will not produce fertile or viable offspring.

This concept of speciation in terms of reproductive isolation has much to recommend it. Studies of groups of closely related species, which seem to have evolved from common ancestors relatively recently, show that they are likely to differ most from one another in those aspects of their form or behaviour that govern courtship and choice of mate. There are species of frog that seem to be identical in almost every way but for the mating call of the males and the females' response to it. There are species of spider that can be told apart only from the shape of the male genitalia. When species are caught in the very act of their origin, reproductive isolation is the first thing to be established. Only once this is secure can other aspects of a species' form and behaviour begin to vary.

Mayr's biological species concept is an operational one. It sees species in terms of how they behave rather than what they look like. It is Darwinian in that it defines species in terms of the *process* of evolution rather than its *pattern*: in terms of *becoming* but not of *being*. This is a perfectly acceptable and practical way to look at species in terms of processes that we can observe in everyday time. We can watch animals and plants mate and reproduce; we can watch – and even engineer – instances of reproductive isolation in controlled conditions.

Mayr's definition falters in cases in which we know little about the biology – the ecological context – of the species we are looking at. When presented with a pair of identical animals, hitherto unknown to science and outside their natural surroundings, it might be hard to decide whether they belong to the same species or are members of two distinct but similar species that differ only, say, in the mating call of the male. In this case, it may still be possible to travel to these animals' native habitat and discover the details of their habits in the wild.

But Mayr's concept breaks down altogether when confronted with fossils. As we have seen, fossils are often of creatures unlike anything known today, comprehended imperfectly in the light of contingent modern models, and almost always divorced from their contemporary ecological contexts. There is no chance that we will ever be able to observe the living animals in the wild. Nobody will ever hear the mating call of a *Triceratops*, or really know if the several different species of *Triceratops* that have been described actually refer to distinct species according to Mayr's definition (for the term *Triceratops* is technically a genus, meaning a group of different but related species). We can never define species of *Triceratops* in terms of their behaviour; all we have to go on is what they look like. This is the antithesis of the biological species concept, which defines species by their behaviour, not their appearance. The biological species concept may be a practical way to look at species in terms of processes that we can observe in everyday time but it is startlingly inappropriate for looking at the relationships of fossil species in Deep Time.

One palaeontologist, George Gaylord Simpson, a contemporary of Mayr and Dobzhansky, sought to address the problem of what was meant by a species when the element of time is introduced. Should a species be an easily identifiable group, defined on its present-day behaviour, as in Mayr's biological species concept; or should it acknowledge its history? Should a species, therefore, represent the sum of groups over time – a lineage? Simpson chose history and lineage and, in 1961, put forward his 'evolutionary species concept': 'an evolutionary species is a lineage (an ancestral–descendant sequence of populations) evolving separately from others and with its own unitary evolutionary role and tendencies'.[60]

Species are defined on the basis of their ancestry, or, rather, on their position in a sequence of species arranged into ancestors and descendants. This is quite reasonable from the Darwinian perspective, in which species ought properly to be considered as segments of an evolving lineage.

Although it is fine in theory to regard species in this way, it is easy to see why this definition has no practical value because it depends on our being able to take a series of isolated fossils, arrange them into sequences of ancestry and descent, and explain such sequences in terms of adaptive scenarios, narratives in which natural selection is proposed to have acted in a certain way over the course of millions of years: Deep Time. But we know that it is impossible, when confronted with a fossil, to be certain whether it is your ancestor or the ancestor of anything else, even another fossil. We also know that adaptive scenarios are simply justifications for particular arrangements of fossils, made after the fact, and which rely for their justification on authority rather than on testable hypotheses. Finally, we also have good reason to suspect that to use natural selection to explain long-term trends in the fossil record may not be a valid exercise, because natural selection is a random, undirected process, unlikely to work in the same direction for long. Ultimately, any trends we see in Deep Time are recognised by us, after the fact.

Simpson's evolutionary species concept was an effort to codify the views of palaeontologists trying to fit their discipline into the Modern Synthesis of evolution and genetics by bringing Darwin up to date and explaining the history of life in terms of adaptation, natural selection and genetics. As long ago as 1933, Simpson's colleague, the distinguished palaeontologist Alfred Sherwood Romer, published an evolutionary scenario that serves us here as a case history of this kind of thinking in action.[61]

Romer wondered why many kinds of primitive and ancient fishes, such as the jawless pteraspids we met in Chapter 2, were often heavily armoured. He observed that in rocks of Silurian age, armoured fishes of all kinds tend to be accompanied by fossils of creatures called eurypterids. Eurypterids were aquatic relatives of scorpions. Like scorpions, they were segmented animals with jointed legs and some species had formidable-looking sets of pincers. Some eurypterids grew to three metres in length. Eurypterid fossils become less common in the succeeding Devonian Period and this fall is accompanied by a decline in the species of fishes with heavy body armour.

Romer speculated that there might have been a causal connection between the presence of eurypterids and armoured fishes. Specifically, armour could have evolved as a defence against eurypterid attack. As eurypterids declined in numbers during the Devonian, the selective pressure on fishes to maintain their armour was weakened, leading to the appearance of more and more species of less well-armoured fishes. Fishes without armour would have been better, faster swimmers than those encumbered by bony shielding. Perhaps fancifully, Romer wondered whether the eurypterid decline might even have been accelerated by the appearance of unarmoured, fast-moving predatory fishes that chased and hunted down the last of the eurypterids: the prey had become the predator.

Although this explanation might be right, it can never be justified. It paints a cartoon version of reality, in which all that matters are the jousting vertebrates and eurypterids, playing down what might have been a complex ecological interrelationship and ignoring other possibilities. It could have been, for example, that the abundance of both armoured fishes and eurypterids was controlled by some other independent factor, such as the abundance of some food resource on which they both depended. It could have been, for example, that bone evolved for a purpose other than defence. Bone is made from calcium phosphate; the evolution of bony armour could have been a convenient way of hoarding phosphate, which is vital for life but scarce in the sea. Deflecting the unwelcome attentions of predators would have been an incidental benefit of this armour.

The scenario also makes light of the timescale in which it takes place. The Silurian Period was thirty million years long; the Devonian, forty-eight million: Romer's tale is truly an epic, unfolding against a backdrop of seventy-eight million years. Yet Romer takes instances of fossils from these immense geological periods and tells a connected narrative of the acquisition and loss of armour, processes that presumably took millions of years of unwavering, directional natural selection to achieve. 'While all known early vertebrates were superficially well armored', he says, 'a rapid increase in unarmored forms and a strong tendency for the thinning out and loss of dermal defenses appears to have been initiated during the Devonian'.

Words such as 'rapid', 'increase', 'tendency', 'thinning' and 'loss' are words that imply the passage of time. The phrase 'decline and fall', used elsewhere in Romer's article to describe the gradual disappearance of the eurypterids from the fossil record, is an allusion to Gibbon's *Decline and Fall of the Roman Empire*, a title that historian Niall Ferguson sees as an exemplar of a style of history told according to the conventions of narrative form, with a beginning, a middle and an end, key characters, key events, and a linear plot.[62] Romer's use of such words and phrases suggests that he thinks that evolution can be recounted as if it were a human drama. Once evolution can be told according to a comprehensible scale of time – in other words, on a scale of time that admits connected narrative – it is a small step to take the evidence from individual fossils and use them to recount a story of evolution that implies knowable ancestry and descent. The ancestors were armoured, but a 'strong tendency for the thinning out and loss of dermal defenses' led to unarmoured descendants.

The story has not only a plot but a motive, in the form of adaptive purpose. The terms 'armor' and 'dermal defenses' are loaded; they show that Romer had prejudged the conclusion of his scenario. These terms imply attack and predation but these implications are speculative because they cannot be tested. The evolution of bony armour as a phosphate reservoir, though equally plausible as an adaptive purpose, makes a far less exciting tale – certainly much less dramatic than epics about ancient wars.

Finally, Romer's tale is a medieval romance: a chronicle of battles in a romantically inaccessible past, in which our brave little ancestors, outnumbered and with inferior armaments, conquered the huge, nightmarish adversary. The story seems plausible not because of its scientific rigour but because of its emotional resonance. We are expected to sympathise with those beleaguered little fishes and feel revulsion at the clattering pincers and glassy stares of the eurypterids. Romer was one of the finest writers that vertebrate palaeontology has ever produced – even his textbooks are a joy to read – so his skill as a storyteller may be all you need to be convinced that his story is true. That, and his authority, reinforced by the appearance of his scenario in a respected and widely read scientific journal.

Romer and Simpson were arguably the most influential vertebrate palaeontologists of the period immediately after World War II. Together, in books and articles, they presented a strongly Darwinian picture of evolution, interpreting structures in terms of presumed adaptation, extrapolating the activities of natural selection over millions of years, and presenting it in terms of accessible stories such as the tale of the eurypterids and the armoured fishes. The Romer–Simpson view was very much the establishment view of evolutionary history.

The popular view of evolution as progressive and inevitable – ad-man's evolution – is rather different from the Romer–Simpson view. Instead, the popular view can be considered as a kind of hybrid of the Darwinian with the classical view. I hinted that Darwin's views, although antithetical to this older view, did not completely supplant it.

Now I can make my hints more explicit. Although Darwinian natural selection gradually became the accepted world-view of people interested in natural history and, later on, experimental genetics, it had less success, at least initially, in palaeontology. As a result, palaeontology as an intellectual pursuit languished for almost a century, between the *Origin* and the Modern Synthesis. Simpson found palaeontology in a bad way and sought to rehabilitate it. The evolutionary species concept was the result.

The generations of palaeontologists immediately after Darwin – the generations berated by Robert T. Bakker (the man who reinvented *Triceratops* as a giant galloping rhino that laid eggs) as inferior intellects, demoting dinosaurs from fighting cocks to lounging lizards – took natural selection and simply grafted it on to the older, classical plan. The Great Chain of Being was not abolished but became animated by natural selection, which was seen as a force that drove forms from archetype to archetype in predetermined ways. Such a hybrid creed, seeking to merge two antithetical systems, one with archetypes, the other without, could only ever be a caricature. In the analogy of the Periodic Table, it is like proposing an evolutionary trajectory down the alkali metals from lithium to caesium. This is clearly nonsense. The espousal of the Modern Synthesis by Romer and Simpson reflects a reaction to this kind of muddled thinking.

Before Simpson and his colleagues got to work on it, voodoo palaeontology threw up all kinds of interesting and pathological phenomena, such as the idea of 'racial old age' (racial senescence) in which species, like people, had set lifespans after which they would become extinct, to be supplanted by superior, more 'adapted' forms. The dinosaurs were thought to have fallen victim to this, explaining Mike Benton's finding, described at the beginning of this chapter, that the subject of dinosaur extinction attracted little specific interest for a long time. The dinosaurs would have become extinct when their number came up, as part of the natural order of things – there was nothing to explain. Naturally, the superior mammals would have taken their place, as night follows day.

The idea of racial senescence was supported by evidence which we would nowadays think of as rather quaint, if not actually pathetic. Ageing species, like ageing people, were thought to become progressively more prone to malady and mishap. In the same way that older people are more prone to cancer, older species tended to produce pathological forms, the evolutionary equivalent of tumours. Take, as one example, the ammonites. These relatives of the squid and octopus lived in coiled shells. For hundreds of millions of years, these shells were generally variants on a flat spiral, like a curled ram's horn. Like the dinosaurs, ammonites became extinct at the end of the Cretaceous Period, sixty-five million years ago. Towards the end of the Cretaceous, several forms of ammonite appeared whose shells deviated markedly from the standard ram's-horn shape. Suddenly, it seemed, there were ammonites whose shells were tall and conical; or thrown into extravagant loops, like trombones; or even completely irregular. These forms were seen as symptoms of decadence, in view of which the extinction of the ammonites at the end of the Cretaceous was not surprising.

Because this hybrid palaeontology had grafted the idea of change onto archetypes, evolution was thought to follow pre-set courses from more primitive to more advanced organisms. Natural selection was seen as the motivating force that drove organisms towards their destinies. Not only is this view of life a perversion of the idea of natural selection, it is unsustainable given what we know of the character of Deep Time. Yet such views were taken seriously until Simpson and Romer came along to clean up the intellectual shambles that palaeontology had become by the early twentieth century.

The Origin of Species attained its first half-century in 1909. To commemorate the event, the Linnean Society of London, a respected academic body, held a discussion over two sessions on the origin of the vertebrates – backboned animals such as you and me, *Triceratops* and Fred the cat, *Acanthostega* and pteraspids. This sounds like an excellent theme for a session to commemorate Darwin; after all, few can doubt the inherent interest in understanding the origin of the group of the animal kingdom that includes ourselves.

However, the place occupied by the vertebrates in the tree of life has been – and still remains – surprisingly hard to find. Virtually every group of non-vertebrate animal has been proposed as the closest relative, or direct ancestor of the vertebrates.[63] Our closest kin have been sought among the molluscs and the insects, the crustaceans, the echinoderms and various kinds of worm. As we saw in Chapter 2, the current consensus is that the closest relatives of the vertebrates in the invertebrate world are the sea-squirts and lancelets. This view dates back to the 1860s, when it was realised that sea-squirts and lancelets, then conventionally thought to have been molluscs, displayed features in their embryology and anatomy closely akin to that of vertebrates.

By the time of the *Origin*'s Golden Jubilee in 1909, the consensus was becoming more widely accepted. But the focus of the Linnean meeting was not to welcome new views, but to bury the old. In the debate, a scientist named W. H. Gaskell was asked to defend his views that vertebrates evolved from jointed-legged creatures similar to crustaceans, or perhaps the horseshoe crab, *Limulus* – an animal which has not changed in its basic form for hundreds of millions of years. The differences in structure between vertebrates and crustaceans seem so vast as to be unbridgeable, yet Gaskell was forced to link the two by his own conviction that evolution was progressive, a Great Chain of Being animated by natural selection. The unlikely alliance between these two disparate forms reflected the uneasy union between Darwinian ideas of dynamic evolution and older, classical views of static creation.

Gaskell's efforts to link vertebrates with crustaceans required heroic feats of anatomical imagination, as one form was morphed into the other. Crustaceans and vertebrates are opposites in just about every way: crustaceans have an external skeleton; that of vertebrates is internal. The central nervous system of crustaceans is based on a spinal cord that runs along the belly; in vertebrates, it runs along the back. The eyes of crustaceans develop from the external cuticle; the eyes of vertebrates develop internally, as outpocketings from the brain, and so on. Gaskell had to reconcile all these differences.

This is how he did it: the digestive system of crustaceans consists of a bag-like stomach, connected by a short tube forwards to the mouth, and extending backwards into a long, hollow intestine. Gaskell proposed that the digestive system of a crustacean-like animal had been transformed into the vertebrate nervous system which, uniquely among animals, is hollow and tubular. Even the brain is constructed around central spaces, called ventricles. The crustacean stomach would have become the vertebrate brain and the crustacean intestine would have become the vertebrate spinal cord. The mouth of the crustacean would have turned into the nasohypophysial duct, the nostril-like canal found in primitive fishes such as lampreys and pteraspids, which connects the base of the brain with the outside world. The old gut having been turned into the nervous system, a new gut would have formed underneath, by the fusion of the tips of the legs. The gaps between the knees would have become gill slits. The result of all this rewiring and replumbing would have been an animal with a hollow nerve cord, situated above the gut – a situation seen in vertebrates, and nowhere else in the animal world.

The tragedy is not that this scheme is implausible but that it is based on a perverse reading of evolution as progressive and constrained to follow a predetermined course. Gaskell started with the observation that human beings had the largest, most developed brains of all animals. If, in addition, human beings represented the ultimate achievement of evolution, and this superiority was most obviously manifested in the brain, then evolution should be interpreted as a story of progressive improvement in the nervous system.

Man, the pinnacle of evolution, arose from mammals, described by Gaskell in his presentation as 'the highest race in Tertiary times'. Mammals came from reptiles, the highest form of creation in the Mesozoic Era, which in turn came from the amphibians which Gaskell describes as 'the lords of the Carboniferous epoch', and so on, down to the fishes, such as lampreys, living in the Devonian Period. By extension, fishes had to have evolved from those creatures living in the preceding Silurian Period which had the most complex nervous systems, creatures such as the horseshoe crab.[64]

It seems clear from this that central to Gaskell's thought was a view of evolution that was fundamentally progressive, represented by an improvement in the quality and complexity of the nervous system in successive dynasties of animal, each an advance over the last, until evolution reached its apotheosis in Man. In this view, to propose a close link between vertebrates and, say, sea-squirts or lancelets is illogical, as these are simple, essentially brainless creatures. In the arena of natural selection, reasoned Gaskell, a soft-bodied, stupid sea-squirt or a lancelet would not have stood a chance against an armoured, brainy animal like a horseshoe crab. Constrained by his prejudices, therefore, Gaskell had no choice but to find some way of making fishes out of jointed-limbed creatures such as crustaceans, no matter how elaborate the details required to make that transformation work, and no matter how outlandish the final scheme.

Gaskell's thesis makes a fine example of voodoo palaeontology, in which Darwinian evolution is tacked onto an older view based on archetypes. Gaskell's scheme is based on a form of the Great Chain of Being in which the affinities between archetypes are judged on the basis of the complexity of the nervous system. In itself, this is no worse an arrangement for the natural world than is the Periodic Table, in which the elements are ordered according to the numbers of protons in the nuclei of their atoms. But Gaskell then draws arrows between the archetypes, to produce a scheme of directed evolution whose major feature is a progressive increase in brain size. If humanity is the acme of evolution, then the path of evolution must always lead to humanity; it is constrained to follow the pattern laid out in Gaskell's pre-existing scheme, in which crustacean-like creatures evolve into fishes, which evolve into amphibians, and so on, to humanity. Natural selection is brought in as a motor, driving the increase in neuronal complexity: Gaskell proposes a War of the Brains, in which superior brainpower always wins, because it has greater adaptive value. Such is Gaskell's adaptive scenario, proposed after the fact to justify his view of life.

Although Gaskell's view of the world seems antique, bizarre, even silly, it is really no different from the view of evolution promoted by advertising copywriters, a mongrel of Darwinian and classical views. Darwin would never have said that mankind evolved from the other living apes; he would have had the apes and humans evolving separately from a common ancestor, with each of us adapted as well as we could be to our present circumstances, one climbing trees and eating bananas, the other typing at a computer and eating banana sandwiches. However, deep down, I guess that most of us think that we really did evolve from apes and that we, coming later, are somehow 'improved' or 'better' than they. Between us and the apes stand a series of fossil forms, each occupying its own preordained place in human evolution. If that were not so, we would never be able to hail any link as 'missing', or think of the latest model of car, marketed as a technological improvement over the previous model, as 'Evolved'.

Such connotations do not in themselves invalidate the idea of evolution as progressive. Progressive evolution is sunk by its inherent confusion, created by a forced marriage between a system that relies on archetypes and a system in which archetypes are forbidden. It fails because it tries, in one view of life, to mix old and new views of the world, to mix the *being* of archetypes with the *becoming* of evolution. Evolution by copywriter may be effective at selling cars and beer, but it is totally inadequate as a way to understand or explain the history of life. It is a story, created by our own prejudices, draped over the isolated fossils that come down to us from Deep Time.

So much for the ad-men. But is the Romer–Simpson view of evolution any better? It is, after all, explicitly framed according to Darwinian principles, unmixed by other views of the world's workings. But this is the source of the problem. Darwinian principles rely on a process, natural selection, which can be observed only in the here-and-now. Mayr's biological species concept defines species in terms of process – by what organisms do, rather than what they are. Simpson's evolutionary species concept also defines species in terms of process, this time historical – by lineage, not by appearance. But we cannot test hypotheses about adaptation or natural selection if the subjects are all dead. Mayr's biological species concept is useless if the subjects all lived in the Cretaceous Period or indeed at any time apart from the present. Simpson's evolutionary species concept is useless if, as we have found, you cannot arrange fossils in a line and justifiably assert that they represent a continuum of ancestry and descent. The problem is that Darwinism is dynamic. It is about change, not stasis; about process, not pattern; about tales, not tableaux; about *becoming*, not *being*.

Evolution by natural selection explains much about the world around us. It would indeed be remarkable if it had not shaped the history of life. But our understanding of that history cannot be advanced by creating adaptive scenarios framed according to present circumstances and extrapolating them backwards into Deep Time to make a continuous story. As we know, Deep Time is not a movie but a box of miscellaneous, unlabelled snapshots.

Given such confusion, one can only long for the eighteenth century, when Linnaeus could classify animals and plants without worrying about evolution. The archetypes of the classical world indicated a state of pure *being*, untroubled by process or change. To a palaeontologist, this view of the world has a great deal to be said in its favour. Fossils do not *do* anything, they just *are*. It makes sense to look at them in that light – for what they are – without feeling obliged to fit them in to a preconceived idea of adaptive scenarios, or forcing them to play the part of missing links in a sequence of ancestors and descendants created after the fact.

We do not, however, live in the eighteenth century; we live in the present, in a world that Darwin created. Evolution has happened; species do transmute into other species, and we must find some way to incorporate that fact into our evolutionary view of the history of life.

The answer is to find a way of understanding the history of life that acknowledges being and becoming, pattern and process, ordinary time and Deep Time, but keeps them separate. To return to my advice about the unicorn, we must understand the unicorn's place in the history of life before we seek to know how it got there, and why.

That answer is cladistics, in which organisms are considered very simply, in terms of testable hypotheses about the place they occupy in the pattern that evolution, by whatever means and in whatever circumstances, has created. Cladistics does not see organisms as products of ancestry and descent, the results of a process which we cannot evaluate or test, especially when the events concerned are irrecoverably lost in Deep Time and accessible only to our imaginations – which, being human, are biased towards self-glorification. Because it makes no assumptions about ancestry and descent – causes and effects – cladistics is particularly well suited to palaeontology, the study of evolution in Deep Time. It has turned palaeontology from a book of children's stories into a true science. Cladistics represents a revolution in thought as profound as that of Darwinian evolution by natural selection.

5 The Gang of Four

Centuries of centuries and only in the present do things happen.

Jorge Luis Borges, *The Garden of Forking Paths*

Every lunchtime, scientists from the Natural History Museum in London emerge, blinking, from their laboratories, to cross the surging six-lane artery of the Cromwell Road, and fan out all over South Kensington, each to his or her own favourite bar. Close by is the Norfolk Hotel with its basement wine bar and fancy food, and the friendly Hoop and Toy for those with heartier, more basic tastes. Cheerful Italian restaurants cluster around South Kensington Station. Further away is the Anglesea, in a peaceful street of stately Georgian town houses. Further still, down Onslow Square and round a corner into the Old Brompton Road, is – or was – a pub called the Cranley.

A relic from an earlier age, the Cranley made no pretensions to Kensington chic. A jukebox and a one-armed bandit, its dreary lights flashing into the cheerless interior, substituted for atmosphere. In the early 1980s, a shadowy table at the back of the bar was usually occupied by a gang of regular lunchtime drinkers from the Museum – the Fossil Fish Section, from the Department of Palaeontology. Each lunchtime, the head of the Fossil Fish Section, Colin Patterson, and his colleague Brian Gardiner from the University of London, would drink a rich mixture of draught Best Bitter and bottled Strong Pale Ale. Over the course of a lunchbreak they would match each other pint for pint, filling the table with empties. They would be aided by Peter Forey – Patterson's

Museum colleague and Gardiner's ex-student – and a coterie of Museum scientists, technicians and students.

Everyone else at the Museum knew that the Cranley was the Fossil Fish Section's favourite haunt and they kept out of the way. Besides, to most people, the Cranley was just that bit too far away from the Museum for a swift pint, a fact which ensured that the Fossil Fish Section had the place mostly to itself. To the scientific community at large, the Fossil Fish Section was synonymous with cladistics. The Cranley was 'The Cladist's Arms' – a den of subversives intent on fomenting academic revolution. The cladists were engaged in a guerilla war with the establishment of evolutionary biology, which saw the world in much the same way that Romer and Simpson had done – in terms of sequences of ancestry and descent, from which could be discerned evolutionary trends, bolstered after the fact by adaptive scenarios.

As well as fighting their battles with the establishment, the cladists enjoyed the same licence as any group perceived as somewhat left-field. They tended to keep their own company and relished any opportunity to poke fun at pomposity while shying away from black-tie dinners, conference banquets and other social functions. The establishment viewed this informality with suspicion. But what irritated the establishment more than anything was that the cladists seemed to be enjoying themselves at everyone else's expense, and nobody likes being made a fool of. The cladists were Mad, Bad and Dangerous To Know. To an impressionable student like me, this combination exerted a magnetic attraction.

I got to know the Cranley in July 1983, when I came down to London to learn about pteraspid fishes and their head shields. My route was a circuitous one. When the Natural History Museum closed the Hall of Fossil Fishes of my boyhood, I moved on to other things and pretended that I had half-fallen out of love with fossils. As I grew older, I thought palaeontology was something for kids, not careers – so in 1981 I went up to Leeds University to read a joint honours degree in biochemistry (for job prospects) and zoology (for fun).

My reconciliation with fossils – the decision to become a palaeontologist – came on me suddenly, in the autumn of 1982, when I was interviewed by the Professor of Zoology, R. McNeill Alexander. McNeill Alexander is the world's leading authority on the mechanics of animal movement. Every year, he took time from his busy schedule to interview the several dozen second-year students reading zoology, either as a sole or joint major, so he could get to know them all in person.

McNeill Alexander is a tall, kindly man with a long, white beard, who radiates charm and good humour. To an anxious second-year student, ushered into his presence, he bore a strong resemblance to God. 'What do you want to do with your life, Henry?' asked the Professor, trying to dispel my obvious nerves.

'I want to be a vertebrate palaeontologist', I replied, somewhat to my own surprise. The words just popped out. At that point, McNeill Alexander bent over to me and said, as gently as possible, 'You do realise that there aren't many *openings* in vertebrate palaeontology?' At that point, he told me of the Natural History Museum's summer studentship programme, and encouraged me to apply.

When the details of the 1983 summer programme arrived, McNeill Alexander and I sat down to choose which of the several projects on offer would be best for me, given that applications were competitive (and that the successful applicant would take home the generous sum of £104 a week). I could have chosen to rearrange the ichthyosaurs, or dust the *Triceratops*. These projects, or indeed anything involving dinosaurs, sounded too glamorous and were very likely oversubscribed. In the end, we selected a project run by the Fossil Fish Section described by just two words, 'pteraspid fishes'. There were two significant advantages; first, I knew what pteraspid fishes were, having been one of the few people ever to have visited the long-vanished Hall of Fossil Fishes. Second, the project seemed so obscure that nobody else was likely to have applied. I got the job. Indeed, as Peter Forey told me later, I had been the only applicant.

On my first day I learned never to underestimate the Old Boy Network. 'You're from McNeill Alexander's department at Leeds, aren't you?' asked Colin Patterson, in the kind of resonant voice rarely heard outside productions of Shakespeare. 'He was in my form at school. Clever sod'.

That was all I heard from Patterson for two weeks. He was off to New York that day, to work with a colleague called Donn Rosen at the American Museum of Natural History. In the meantime, Forey set me to work on pteraspid fishes. A man of few words and an inordinate fondness for Carlsberg Special Brew, Forey would let me get on with it while he pored over fossil lobe-finned fishes. His colleagues, Sally Young and Alison Longbottom, arranged for me to have a desk at the farthest end of the galleries, so I could work right in the middle of my childhood friends, the fossil fishes.

While I was working away in the collections, surrounded by drawers and boxes of fossils, I would receive distinguished visitors. I thought they were just being friendly. Later I realised that they were scouting for likely graduate students. My first visitor was Patterson's drinking partner, Brian Gardiner. In contrast to Patterson's assured, patrician accent, Gardiner has never quite lost the tones of his native Gloucestershire. He would sooner chat over a pint about his grandmother's recipe for lamprey pie, or his almost obsessional interest in the famous Piltdown Man hoax of 1912,[65] as talk about the fossil fishes he's described in monograph after monograph. To do otherwise would seem immodest; after all, the monographs speak for themselves. Gardiner, Patterson and Forey took me out to the Cranley every lunchtime and bought me lots of beer. However, it took me a little while before I realised what was going on; why the scientists in the Fossil Fish Section kept their own company, and why other scientists stayed away from the Cranley as if there was a curse on it.[66]

The reason was a 112-page technical monograph entitled 'Lungfishes, Tetrapods, Paleontology and Plesiomorphy', written by Donn Rosen, Peter Forey, Colin Patterson and Brian Gardiner, published in 1981 as an article in the scholarly *Bulletin of the American Museum of Natural History*. In the article, the authors argued that lungfishes (those curious, semi-amphibious fishes found in South America, Africa and Australia) were the closest living relatives of the tetrapods.[67]

At first glance, such a conclusion seems quite uncontroversial. However, the monograph was not written in the cool, measured language one associates with a scientific paper. As well as being unashamedly technical (even the title is hard work), it was uncompromisingly polemical in tone, bristling against the Romer–Simpson tradition of ancestor–descendant relationships.

Second, the paper argued that the relationship between lungfishes and tetrapods could be inferred, using cladistics, from the living forms alone without reference to fossils. Rosen and colleagues claimed that a century of scholarship had been wasted trying to draw sequences of ancestry and descent between fossil lobe-finned fishes, such as *Eusthenopteron*, and extant tetrapods. Imagine that you are a traditionally minded palaeontologist who has just read this paper. To be told that your calling is of no significance, having been built on a century of futility, is bad enough; to have this disheartening message emphasised in the strong language employed by the authors is worse; but that the paper had been written by *four of your fellow professionals* was quite beyond the pale. Palaeontology is a discipline perpetually short of funds and constantly striving to maintain its credibility. An attack by insiders on the conceptual roots of the discipline might be considered treachery.

As may be imagined, Rosen and colleagues were greeted with the same incredulity as that attending a broker who walks into the stock exchange and declares that money is an illusion brought on by greed. By virtue of their perceived ruthlessness towards the methods espoused by their more traditionally minded colleagues, the authors – Rosen, Forey, Patterson and Gardiner – acquired a collective soubriquet: the Gang of Four.

Much of the substance of the Gang of Four's monograph concerned the anatomy of the snout of tetrapods and lobe-finned fishes, in particular the nostrils. The nasal passages inside your head have two sets of openings. Your nostrils comprise the nasal passages' front gate. But the nasal passages also have rear entrances, situated in the roof of your mouth. These internal nostrils, called 'choanae', are found in all tetrapods. Most fishes, however, do not have choanae – their external nostrils open into blind sacs, which have no connection with the mouth cavity. The exceptions, claimed the Gang of Four, are the lungfishes.

The Gang of Four claimed that the choanae of lungfishes are best seen by dissecting carcasses rather than looking at fossils. The reason is that the choanae are, essentially, soft-tissue structures. When we look at choanae in a freshly dead fish, we can trace every detail of their anatomy. We can insert a probe into an external nostril and watch it emerge from a choana. The identification of lungfish choanae is unambiguous. But claims for choanae in fossil fishes are based on the interpretation not of soft tissue, but of gaps between bones. A choana in a freshly dead lungfish can be identified as such with confidence, but a gap between two bones in a fossil with no close modern model is only a gap. Because we will never be able to study fossil fishes in the flesh, the presence of choanae in fossil fishes cannot be established with the same certainty as with extant lungfishes. The Gang of Four argued that holes in the palates of fossil lobe-finned fishes such as *Eusthenopteron* would have been too narrow, when clothed in flesh, to have made convincing choanae.

The argument went beyond anatomical revisionism. As I described in Chapter 2, convention had represented osteolepiform fishes such as *Eusthenopteron* as our ancestors. When tracing sequences of ancestry and descent, we naturally look for the features in our purported ancestors that mirror our own – after all, *we* are the descendants that get to write the book. We look at tetrapods and we see choanae; coloured by this prejudice, we look back at *Eusthenopteron*, already regarded as our ancestor, and interpret gaps in the palate as choanae, when they might as easily have represented other structures with no close correspondent in modern animals, or might just have been gaps between the abraded bones in imperfectly preserved fossils. The Gang of Four exposed this line of argument as fallacious. Fossils are not real animals: they are imperfectly preserved fragments which we interpret, after the fact, in the light of modern models. It is all too easy to interpret ambiguous structures in fossils to suit pre-existing stories about ancestry and descent. The Gang of Four tried to show that lungfishes, like tetrapods, are choanate, but the same could not be said with the same certainty for fossil lobe-fins.

The possession of choanae suggested that tetrapods and lungfishes share a common ancestry that excludes other fishes, in the same way that the shared possession of pointed ears and whiskers is evidence that my cats Marmite and Fred have a common ancestry that excludes me.[68]

This arrangement has a logical consequence: if some fishes are regarded as more closely related to tetrapods than are other fishes, then it will be impossible to come up with a list of features – fins, scales and so on – that defines fishes to the exclusion of all other animals, including tetrapods. Fishes do not form a natural group. They are simply all those vertebrates that are left over once tetrapods have been taken away. A fish is simply something that you buy from a fishmonger – the term 'fish' has no zoological meaning as a category.

When Patterson and colleagues presented a preliminary version of their findings at a conference at the University of Reading in 1978, it provoked an exasperated response from one delegate, the palaeontologist Rex Parrington from the University of Cambridge. If the cladists were correct, Parrington exclaimed, then a lungfish would be more closely related to a cow than a salmon. The implication – that the term 'fish' was redundant as a concept – was not lost on Parrington, who clearly found it outrageous. The argument can be summarised in two simple cladograms, as in Figure 11.

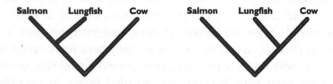

Figure 11. The problem of the Salmon, the Lungfish and the Cow. On the left is the traditional view as espoused by Parrington; on the right is the scheme adopted by the Gang of Four.

Parrington's discomfort, shared by other delegates at the meeting and palaeontologists in the wider world, came from a confusion between pattern and process. In the classical world of archetypes, fishes would be as distinct from tetrapods as the element carbon is from the element nitrogen. Even though in some instances fishes evolved to become tetrapods, the fish archetype would remain.

Although palaeontologists such as Parrington acknowledged the Darwinian, process-based view of evolution as promoted by Romer and Simpson, their thinking was still coloured by the notion of archetypes. This is why Parrington thought it outrageous that the very idea of fishes could become redundant; but this, the abolition of archetypes, was no more than implied by Darwinian evolution. The cladists were simply pointing out the problems of trying to reconcile two irreconcilable systems of understanding the pattern of nature: if some fishes evolved to become tetrapods, then there is no way to define fishes in a way that excluded tetrapods. And what is an archetype but a prescription of features that all fishes ought to share? If we admit the existence of evolution in the Darwinian mode, the cladists claim, then we must be true to the spirit of Darwin and abolish the archetype.

The cladists were making a distinction between two kinds of category, 'clades' and 'grades'. A clade is a branch on a cladogram, representing a group of organisms defined solely on the basis of degrees of cousinhood. The cladogram on the right of Figure 11, in which the lungfish and the cow form a clade at the expense of the salmon, shows that lungfishes and cows are closer cousins than either is with the salmon. This is just like the cladogram in Figure 3, in which my cats Fred and Marmite form a clade at my expense; Fred and Marmite are closer cousins to each other than either is to me. The term 'cat' denotes a clade, because it implies an inventory of features that defines cats uniquely (whiskers, retractile claws, pointed ears, chasing mice and so on). This is the list of features that one would expect to find in the most recent common ancestor of Fred and Marmite.

A grade, in contrast, reflects similarity of structure, which need have no connection with recency of common ancestry, degree of cousinhood, or indeed with any aspect of relationship. Fishes – that is, all vertebrates that are not tetrapods – represent a grade, united by community of aquatic habit rather than community of descent. Fishes are simply all those vertebrates with fins, or a label for all the vertebrates left over once you have taken away all those fishes with legs. This means that although the term 'fish' can be used informally to mean something you buy from a fishmonger, it does not have a precise zoological meaning. In the same way, a term such as 'quadruped' denotes a grade rather than a clade because there is no inventory of features that uniquely defines such four-legged beasts.

It seems clear that grades and clades are two entirely different things. Fundamentally, clades can be defined according to a prescribed list of features shared by all the members of that clade. All cats, for example, have pointed ears, whiskers, retractile claws, a fondness for chasing mice and so on. Taken together, this list offers a unique diagnosis of cats. Grades, on the other hand, can be defined by no unique list. There is no way, for example, to produce a list of features found in all fishes but no other creatures – unless tetrapods are also included, as fishes with legs. Once tetrapods are taken away, fishes just become a loose assemblage of aquatic vertebrates.

In the Romer–Simpson view of life, no distinction was made between grades and clades. Romer and Simpson recognised that although grades need say nothing useful about relationships – that is, degrees of cousinhood – grades reveal much about presumed adaptive transitions in evolution. For example, the term 'fish', as a concept, gave access to a simple mental picture of what the ancestors of tetrapods looked like. However, we now know that ideas about adaptation and ancestry are largely untestable; to think of fishes as ancestors is to run the risk of falling into the same trap that ensnared W. H. Gaskell, who based his idea that we all came from crustacean-like ancestors on a fundamental misreading of evolution as inevitable and progressive. And, as we know, Deep Time is an unsuitable medium for telling tales.

The importance of the distinction between grades and clades was established in 1950, when an East German entomologist called Willi Hennig published a book setting out how to classify organisms strictly on the basis of recency of common ancestry, repudiating criteria of general resemblance. Hennig called his method of classification 'phylogenetic systematics', as it was a form of systematics based on phylogeny – the evolutionary, genealogical pattern that binds organisms together. It was only later on that phylogenetic systematics came to be called 'cladistics'.[69] Although Hennig was not especially interested in fossils, cladistics was found to have particular relevance to palaeontology, as it is a way to describe the pattern of the history of life, free from the problems inherent in ancestor–descendant sequences or adaptive scenarios. Because cladistics describes a pattern rather than tells a linear narrative, it is uniquely suited to the study of Deep Time.

The fundamental, central concept in cladistics is what Hennig called the 'sister-group relationship', in which organisms are grouped together on the basis of their shared common heritage. Classification should not be a search for ancestry and descent, said Hennig, but an identification of the sister-group relationships between organisms – in other words, of degrees of cousinhood. The diagram in Figure 2 – showing how myself and my cat Fred are related through a common ancestor, at the node – is an example of a sister-group relationship. The actual participants, Fred and myself, appear at the tips of the branches, not halfway along the branches or at the node. The common ancestor at the node is an inference only. Were we ever to discover the most recent common ancestor of Fred and myself, we could never know that we had done so.

Such a diagram could be drawn for any pair of organisms, living or extinct. It could apply to two fossil fishes, 'a' and 'b'. Confronted with these two fossils, one might speculate that 'a' (which is geologically the older of the two) is ancestral to 'b'. More broadly, we might say that the species to which 'a' belongs (let us call this species A) is ancestral to the species to which 'b' belongs (let this be species B). We could never know this to be true. Why? First, because fossils are not found with their pedigrees; second, because there is at least one other alternative. That is, it is also possible that species B is ancestral to species A.

How can this be, if fossil 'a' is older than fossil 'b'? It is conceivable that species B could have evolved first. Species A could have evolved from a small population of B, but the main population of B continued alongside A. Given the incompleteness of the fossil record, it is possible that fossil 'a' could be a very early example of species A, and that fossil 'b' is a late example of species B, younger than 'a'. To say, then, that A is ancestral to B might be wrong – but we can never prove the second scenario, either. (The options are explored in Figure 12.)

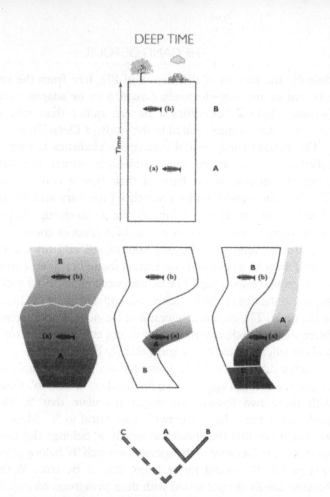

Figure 12. Cladograms and lineages. How one cladogram summarises two or more possible scenarios of ancestry and descent. Top: the geology shows that fossil 'a', a member of species A, is older than fossil 'b', a member of species B. Centre: three possible schemes of ancestry and descent. A could be ancestral to B, B could be ancestral to A, or they could both descend from a third, unknown species, C. On this evidence, all are possible. Bottom: the cladogram summarises all these possibilities but it is solely based on the morphology of the fossils 'a' and 'b' and makes no assumptions about ancestry and descent.

The only solution is a pragmatic one. With no certain knowledge of what actually happened, all one can do is say that species A and B (as represented by fossil 'a' and fossil 'b') form a sister-group relationship. Species A *could* have evolved from B; B *could* have evolved from A; but both *could* have evolved from a common ancestor, which may have been a member of A or B, or another, unknown species, C. We can never know for certain, but why worry? Cladistics makes all this speculation unnecessary by adopting the practical solution – even if we can never know what actually happened, we *do* know that fossils 'a' and 'b' are somehow related through a shared common heritage, because such a relationship *must* be true for *any* pair of organisms. The sister-group relationship is a convenient summary not of what happened, but of all the things that *might* have happened.

In the early 1960s, Hennig's work came to the attention of a young American palaeontologist called Gareth Nelson, who happened to be working on fossil fishes. For Nelson, cladistics shone a bright light onto the murky procedure by which palaeontologists used fossils to infer ancestor–descendant relationships. How can one ever know that a fossil in your hand is your ancestor, or anyone else's ancestor, Nelson asked himself? As Nelson put it, rhetorically, in a lecture in the early 1970s – 'Do the rocks speak?'[70]

Nelson asked himself a simple question about ancestry and descent and found that he was staring Deep Time straight in the face. Because Deep Time admits no link between cause and effect, it is clear that no scheme of ancestry and descent in which known fossil forms are linked together can ever be justified, because we can never know whether the fossils concerned are ancestral to each other or to anything else. As Nelson wrote in the same lecture, 'Looking for ancestors in the fossil record seems to be something like looking for honest men: in theory they must exist, but finding them in practice, alas, is another matter'.

On a visit to London, Nelson shared his new enthusiasm with Colin Patterson, who was also mulling over the implications of Hennig's ideas.[71] Patterson spread the news to his friend Brian Gardiner, who had a student called Peter Forey, who would eventually work with Patterson. Later, when Nelson joined the staff of the American Museum of Natural History in New York, he convinced Donn Rosen that the inference of ancestor–descendant relationships, and thus the basis for palaeontology in the Romer–Simpson mould, was wrong. Rosen and Patterson, both students of fossil fishes, became ardent proponents of what came to be known as cladistics. The Gang of Four was born.

Another visitor to my desk in the remoter parts of the national collection of fossil fishes was the idiosyncratic palaeontologist Lambert Beverly Halstead, universally known as 'Bev'. It was he who had convened the meeting at the University of Reading that led to the framing of the contentious issue of the Salmon, the Lungfish and the Cow – and who had written it up for *Nature*, where it was published on 21 December 1978. His critical treatment of cladistics led to a spirited correspondence in the journal as 1978 turned into 1979. Once an active Marxist, Bev's political opinion had veered away so sharply that he saw cladistics as a communist plot to take over the Natural History Museum and corrupt the young.

On meeting Bev in person, you were never quite sure if his objections were sincere or if he was causing a commotion just for devilment. Bev adored publicity. With his floppy fringe of white hair, gleaming eye, and a private life which might charitably be described as 'Bohemian', he looked less like an academic than a middle-aged rock guitarist, still not above throwing a few metaphorical TVs into swimming pools for the sake of a few column inches. The manner of Bev's death was as shocking and dramatic as his life – the car he was driving collided with a truck and caught fire.

Bev's comments in *Nature* in 1978 marked the first time that arguments about cladistics had been aired outside specialist technical journals on classification, where they had been smouldering acridly for more than a decade. In 1980, Bev fired a second broadside, this time castigating the Natural History Museum for applying cladistics in two new exhibitions: *Dinosaurs and their living relatives*, which opened in 1979, and *Man's place in evolution*, which opened the following year. This letter was published in *Nature* on 20 November 1980 and this time Bev exceeded even himself, equating cladistics with the advocacy of sudden evolutionary leaps (rather than gradual change) and, through that, with communist revolution. If the

Museum had its way, said Halstead, then 'a fundamentally Marxist view of the history of life will have been incorporated into a key element of the educational system of this country'.

Cladistics is concerned with the pattern produced by the evolutionary process; it is not concerned with the process that created the pattern or the swiftness or slowness with which that process acted. Neither is it a political creed, overtly or covertly. However, Bev's letter produced a cauldron of correspondence on a huge variety of topics ranging from museum exhibition policies to the relevance (or not) of political ideology to the study of the history of life. Whenever the correspondence looked like flagging, *Nature*'s leader writers kept the pot boiling with a few playful thunderbolts of their own, such as an accusation that the scientists of the Natural History Museum doubted the truth of evolution.[72] The cauldron bubbled for six months – if Bev wanted publicity, he got it.

When one strips away the confusion about gradual and sudden modes of evolution, not to mention the flummery about Marx and Engels, Halstead's public denunciation of cladistics rested on his Simpsonian views of evolution and the fossil record. On the subject of human evolution, Halstead wrote, 'The well-attested sequence of human fossils representing samples of succeeding populations has, until the Natural History Museum's latest exercise, been taken as a classic example of the gradual evolution of a single gene pool. Certainly there is not any serious doubt about *Homo erectus* being directly ancestral to *Homo sapiens*'.[73]

On 4 December 1980, Colin Patterson replied directly to this passage:

> Confronted with these statements, one must either bow to Halstead's scholarship, or ask 'attested' by whom? 'been taken' by whom? Halstead's answer might be, to quote the Museum's handbook to the old exhibit on fossil man which was removed to make way for the dinosaurs, 'the evolution of man has come to be regarded as fact rather than hypothesis by all persons qualified to judge the evidence'. In other words, we (scientists, experts, authorities) tell you it is so. The radical departure in the exhibit reviled by Halstead is that the voice of authority is less strident. The visitor is encouraged to understand, and to take part in, the

reasoning that underpins the story of human evolution: to become one of those 'persons qualified to judge the evidence'.[74]

On 1 January 1981, Donn Rosen added his own voice to the debate.[75] His riposte to Bev in *Nature* was savagely ironic:

> Halstead states his convictions that human evolution was gradual, that its gene pool of the past had certain knowable characteristics, and that the ancestry of a living species can be determined with deadly accuracy. Halstead's convictions arise from his discovery of a new form of *doubt*, such that he can contend that 'there is not any serious doubt about *Homo erectus* being directly ancestral to *Homo sapiens*'. This is certainly the news Biology has waited for, the moment when the Truth can at last be known so that all this difficult and extremely tiresome theory can be dispensed with. Until now my colleagues and I had always imagined that to doubt something was 'to be uncertain as to a truth or fact' and the notion that distinguishes science from, say, politics is that in science uncertainty about the truth must remain or progress ends. Halstead's discovery can only mean that he has in hand a new form of truth – a kind of truth that can be known. Many of us puzzled over what kind it might be until it finally dawned on us that it emerges from false doubt and, to honour its discoverer, I call it *Halsteadian Truth*. [original emphasis]

He continued: 'Now that Halsteadian Truth permits us to know at last the way in which evolution proceeds, what extinct gene pools were like, and exactly who was whose ancestor, I confess, with sadness for years misspent, that cladistics is indeed a waste of time'.

If, as Bev maintained, there is 'not any serious doubt' that *Homo erectus* is ancestral to *Homo sapiens*, there cannot similarly be 'any serious doubt' that any fossil species can, in principle, be linked up in chains of ancestry and descent with any other. Such links can be proposed, but they exist solely in the mind of the beholder and can never be tested. Why? Because the gap between evidence and model is too great to bridge. The actual evidence consists of skulls and bones, discontinuous and disjunct, isolated from one another as the tableaux that punctuate the endless corridor of Deep Time. But the evolutionary model as promoted by Simpson and Romer, and followed by Halstead, speaks of something qualitatively different. That is, the continuous, seamless evolution of a single gene pool from one species to another – a process that implies narrative, a property of ordinary time that is alien to the element of Deep Time, the home of fossils, not gene pools.

Therefore, one can never, with justification, impose a particular scenario of continuous ancestry and descent on a pile of bones – except by assertion. As Patterson commented, such a fact is said to be true because we scientists claim to be 'persons qualified to judge the evidence', and we tell you that it is *so*. To be able to make such assertions, scientists must be confident that they are in possession of the Truth – in Rosen's words, 'a kind of truth that can be known'.

Rosen's comments illuminate the heart of cladistics and, indeed, of scientific enquiry. From our schooldays, we are taught that science progresses by the framing and testing of hypotheses. More broadly, we are taught to be sceptical, to demand access to the evidence and to be prepared to spend time verifying the assertions of others. We are taught that the credibility of science rests on such institutionalised scepticism. We are taught that the strength of a hypothesis and its accessibility to experiment are of greater value than the perceived eminence of the experimenter; we are taught never to accept the validity of a hypothesis simply because our superiors tell us that it is true. In short, we are told never to take anyone's word for it.

The cladists have always taken this open, questioning spirit of science literally and seriously. The constant scepticism voiced by young researchers gives modern palaeontology, since the mid-1980s very much founded on cladistics, a vibrancy that outsiders see as quarrelsome and disputatious. The generation of the Gang of Four encouraged this. For example, my summer work in the Fossil Fish Section often forced me, a complete beginner, to make decisions about taxonomy; I had to reclassify specimens of pteraspid fishes, renaming them according to my reading of Alain Blieck's thesis. I had to write out new labels and reshuffle the entries for each fossil in the Museum card index. On one occasion I had a crisis of confidence – what right had I, a novice who had done no serious work on fossils, to rearrange the national collection? I took my worries to Peter Forey. 'Don't worry about it', he counselled, 'taxonomy is only a matter of opinion'. The implication was that my opinion counted – it was as valid as the opinion of qualified scientists such as Patterson, Rosen, Gardiner or Forey. To paraphrase Patterson's response to Halstead, I was encouraged to participate in the reasoning that underpins the story of evolution; to become one of those 'persons qualified to judge the evidence'.

This openness is very far from the science portrayed by the media. Scientists are paraded on TV or quoted in print to voice their opinions on various subjects as pundits, as experts, as 'boffins', as 'persons qualified to judge the evidence', as if they were the ordained guardians of secret arcana, imparted sparingly to the gawping masses. The danger for scientists is that they will come to believe the hype – that they are indeed secular priests in possession of the truth with a capital 'T', 'the Truth that can be known'. But such truth is expressedly unscientific.

Scientists in general, and palaeontologists in particular, must be aware of the limitations of evidence. The Romer–Simpson tradition of palaeontology is unscientific because its conclusions are based not on hypotheses that can be tested but on assertion and authority. It is also unscientific because it relies on statements that go beyond the evidence – statements on the position of fossils in sequences of ancestry and descent and the adaptive purpose of structures. Cladistics, in contrast, is acutely sensitive to the nature of evidence and to what can – and cannot – be concluded from it. This is why all cladograms are provisional and why they are based on the sister-group relationship. As explained earlier, the sister-group relationship represents a minimum default statement of relationship. Even though it is impossible to know for certain whether one species is the ancestor of another, we *do* know that any two organisms found on Earth must be cousins in some degree.

The limitations of fossil evidence exposed by cladistics are quite severe because any statement which is, at root, based on presuppositions of ancestry and descent will immediately be suspect. Fundamentally, such presuppositions make the mistaken assumption that Deep Time can support a narrative, which we know it cannot. Many discussions on cause and effect (such as the reasons for extinction of the dinosaurs) and the tempo and mode of evolution (gradual and slow, or rapid and 'punctuated') make tacit assumptions that Deep Time is a continuous medium against which a story can be told, whether of ancestry and descent in particular or of connected events in general. As we have seen, individual events in Deep Time are isolated and disconnected from all other events; any connection we make between them is artificial, made by us after the fact, based only on our prejudices and what we as parochial humans living in the particular, peculiar circumstances of the present day think is plausible or possible. But we have no ticket to eternity, no over-arching view of the entirety of creation past, present and future. As scientists, we would do well to confine ourselves to what is practical and possible and avoid claims that we know the 'Truth'.

Cladistics makes a clear distinction between the pattern of evolution and the processes that created that pattern. To repeat my homily on the unicorn at the end of Chapter 2, cladists would seek to discover the situation occupied by the unicorn in the pattern of life before speculating about *why* and *how* it achieved that position. In making clear the distinction between pattern and process, cladistics acknowledges the difference between being and becoming and recognises – at last – that Deep Time is qualitatively different from ordinary time. In that recognition, the tension between being and becoming is resolved.

This resolution has a further, liberating consequence. Evolutionary scenarios cast on the traditional Romer–Simpson mould mix pattern and process. As a result, they are limited by their particular contexts. For example, a scenario in which fishes evolved limbs for the purpose of walking on land applies only to that particular context, the evolution of tetrapods from fishes. It cannot be applied to other problems, such as the evolution of the horns and neck-frill in the dinosaur, *Triceratops*. Cladistics, in contrast, transcends any particular context and becomes a powerful general method for exploring the pattern of life. It does this, paradoxically, by an inherent acknowledgement of its own limitations.

The reason lies in the concept of the sister-group relationship. As I described above, the sister-group relationship is a kind of minimum default statement of relationship. But because it is only a pattern, denying all assumptions about particular processes, courses or causes of ancestry, descent and adaptive purpose, it applies equally to *any* pair of organisms, irrespective of the parochial circumstances that might apply in each case. Whatever the adaptive purpose of limbs, or how they evolved, *Acanthostega* and *Eusthenopteron* were cousins; whatever the reason for the origin of bone, pteraspids and eurypterids were cousins; whatever the reason for the long nectary of the Madagascar Star orchid, all moths and all orchids are cousins. The sister-group relationship is no more than the graphical expression of a truism – that every organism that was, and is, and will be, is a cousin of every other.

By extending this idea, we can see that the sister-group relationship, as a concept, transcends even organic creation. If we believe the truth of evolution, we can assume that all organisms are related by what Darwin called a community of descent: we are all cousins. The sister-group relationship expresses just that – a pattern of relationship – but says nothing about the processes that might have generated that pattern, processes such as natural selection. The implication is that the sister-group relationship can be used to describe the relationship of any two objects related by a community of descent, irrespective of whether the process linking them is natural selection, artificial selection or some other, perhaps inorganic process. By transcending its context, cladistics has become a universal tool for understanding matters as diverse as the evolution of life and the evolution of languages and literature.[76]

In the remainder of this chapter, I shall look at some instances of how cladistics, by acknowledging its limitations, transcends them: the result being a supple, flexible method of understanding many aspects of the history of the world around us.

In Chapter 3, I showed how the evolution of creatures whose remains we pick up as fossils could have been influenced in a profound way by creatures hardly ever seen in the fossil record – in particular, parasites and agents of disease. In that context I specifically mentioned the roundworms, many of which are parasites, and which are scarcely known as fossils. In March 1998, a team of researchers led by Mark Blaxter of the University of Edinburgh, Scotland, presented the first cladogram of roundworm relationships that was informative enough to reveal details of the evolutionary history of this group. They achieved this feat without a single fossil. They read the history of roundworms from modern species, or, to be precise, their molecules.[77]

The problem with roundworms is that although there are many thousands of different species, they all look very much the same. This means that it is hard to find sufficient informative features, based on their anatomy, to produce a robust cladogram. But every animal is a dark continent under the skin. Beneath the anatomy of every organism is a second, occult anatomy – the anatomy of protein and gene sequences. This anatomy presents a panoply of variation even in the most monotonous worm. We can use this variation to work out evolutionary relationships.

Organisms are made of proteins. Proteins, in turn, are made of chains of small, discrete units called amino acids. The structure and properties of a protein are determined by the order in which the amino acids is assembled. This means that around twenty different varieties of amino acids can be assembled in any order to create an inexhaustible variety of proteins. In general, and because there are no limits on the length of an amino-acid chain, proteins can be built to fulfil very specific purposes.

Instructions for assembling proteins are carried in the genetic material, the nucleic acid, DNA. This, in turn, is made of discrete units called nucleotides. There are only four of these, in contrast to the twenty amino acids commonly found in proteins. A sequence of three nucleotides in DNA represents a single amino acid in a protein. This 'triplet code' means that nucleotides can encode, potentially, 64 different amino acids. As there are only twenty common amino acids, there is plenty of slack left in the system for special instructions. Some amino acids are encoded by more than one triplet. Some triplets do not stand for any particular amino acid, but instead are instructions that mean 'start protein sequence here', or 'stop'. Other triplets do not stand for anything at all.

A chain of meaningful nucleotides – a string of triplets – constitutes a gene, which 'stands for' a meaningful chain of amino acids, a protein. Genetic mutations are the result of damage to the nucleic acid chain in which one or more nucleotides is deleted, inserted or changed. Just like changing the letters in a sentence, altering the nucleotide chain can change the whole 'meaning' of a gene. Mutations can disable a protein. Many, though, do little harm. The redundancy of the genetic code ensures that. Besides, many mutations are found in large regions of DNA that do not contain protein-coding genes anyway. They accumulate, unseen, like junk in an attic.

But mutations are the stuff of evolutionary change and tend to accumulate over time as species progressively diverge from their common ancestors. Differences between amino-acid or nucleotide sequences between species alive today are just like any other anatomical differences and their comparison can be used to discern the branching patterns of evolutionary history; the principle of the sister-group relationship applies as much to molecules as to morphology. Cladistic analysis of molecular sequence information, from proteins and genes, has provided a valuable extension to our knowledge and provides an independent foil – a test – for analyses based on morphological information alone.

In the absence of clear physical differences between the various roundworm species, the traditional evolutionary family tree of roundworms was based more on what they did and where they lived. To base an evolutionary tree on the habits of a creature – such as whether or not it is parasitic, or lives on land or in the sea – is fair enough, under the circumstances. This is no help, however, if you want to investigate, for example, the number of times that parasitism evolved independently in roundworm history. If the habit of parasitism is used to presuppose evolutionary relationship, then the relationship cannot be used to investigate the incidence of the habit for to do so would be to follow a circular argument.

Blaxter and his colleagues broke the deadlock, penetrating the relatively uninformative roundworm exterior to reach the rich variety of genetic information beneath. They used gene sequences from 53 species of roundworm to create a cladogram. Their cladogram is useful because the gene sequences used to create it are independent of the habits and habitats of the roundworms they represent. The cladogram can therefore be used to test scenarios of roundworm evolution.

One such commonly held hypothesis is that parasitism evolved just once in the history of roundworms so that all parasitic roundworms form a clade. In contrast, Blaxter and colleagues' cladogram shows that parasitism evolved independently on at least seven occasions, each instance in a distinct group of otherwise free-living roundworm species. It did not happen that, one day, a roundworm swimming freely in a stream found itself within a fish or a snail and parasitism was born.

The discovery that many parasitic species of roundworm have close free-living relatives – rather than forming a distinct clade of parasitic animals – could be important for medicine and agriculture. Parasitologists need to be able to investigate their subjects in laboratory conditions, but it is often difficult or impossible to 'culture' a parasite away from its host. But if a parasite has a close free-living relative, the relative can be used instead as a laboratory 'model' for the parasite.

For example, several agriculturally important roundworm parasites of animals turn out to be close relatives of an innocuous free-living species called *Caenorhabditis elegans*. This species is easy and cheap to keep in laboratories, and happens to be one of the best known of all animals – the complete gene sequence, or 'genome', of the entire animal is now known.[78] Detailed knowledge of the genetics and metabolism of *Caenorhabditis elegans* might shed light on its parasitic cousins, allowing researchers to test the efficacy of anti-parasitic drugs.

This is, of course, all in the future – but the very possibility of such a strategy might have passed people by had Blaxter and colleagues not been able to apply cladistics to gene sequences to determine the evolutionary relationships of the group, in the absence of fossils and stratigraphy, and using living animals alone.

Similar application of cladistics to molecular sequence data is already showing potential to resolve important medical problems. One is the origin of the human immunodeficiency virus HIV-1, which causes AIDS. Viruses, like every other organism, have gene sequences. But viruses evolve at tremendous speed; what would take most organisms millions of years can be achieved by viruses in a few decades or even within the course of a single infection in one individual. This is why viruses are so hard to fight; a vaccination against a strain of influenza this year might be ineffective next year because subsequent strains of the virus have effectively become new species which the vaccine is powerless to combat. This feature has provided researchers with a feast of viral gene sequences, allowing them to track the viruses as they evolve and plot the course of their spread in human populations.

Intensive research on the HIV-1 virus has shown how most isolates of the virus can be classified as belonging to several distinct 'substrains', each one a distinct lineage in the evolutionary history of the virus. Cladistic analyses based on the gene sequences of the various viral substrains can be used to plot how they evolved from a common ancestor. But where did this common ancestor come from – and when? Answers to this question are vital for understanding the history and possible future course of the AIDS epidemic.

Until recently, the first known case was believed to have been a sailor from Manchester, England, who died of an AIDS-like illness in 1959. This case has not been authenticated. In the meantime, results from blood samples taken from people in Africa as long ago as 1959 have come to light. One of these, taken in that same year from a man living in what was then Léopoldville in the Belgian Congo, turned out to test positive for HIV-1. The gene sequence of this virus proved to be archaic by comparison with modern viruses. When plotted in a molecular phylogeny (created by a number of methods used for molecular information, not just cladistics), the Léopoldville virus nestled somewhere near the common ancestor of all extant HIV-1 substrains. In their analysis of the sequence, AIDS researcher David Ho of the Aaron Diamond AIDS Research Center in New York and his colleagues used this phylogenetic knowledge to suggest that the virus entered the human population in Central Africa not long before 1959. Being so close to the common ancestor of all HIV-1 substrains, the Léopoldville virus was probably close to the original source of the epidemic in time and space.[79] More recent molecular sequence comparisons, by Beatrice H. Hahn of the University of Alabama at Birmingham, Alabama and colleagues, has followed HIV-1 back to its original source, among primates. Using molecular sequence comparison, they have traced the origin of HIV-1 to a virus found in a subspecies of the common chimpanzee

(*Pan troglodytes troglodytes*), whose geographic range coincides with the most divergent – and thus most ancient – lineages of HIV-1 in humans.[80]

When applied to molecular sequence information, cladistics can be used to test scenarios about how things came to be – as surely as if the information had come from a comparison of the bones, scales or fins of fossil fishes. Turning to living fishes, anyone with a tank of tropical fish will be familiar with small, brightly coloured fish called swordtails (*Xiphophorus*). In some swordtail species, the lower margin of the tail-fin of the male is extended into a long, pointed 'sword'.

Darwin believed that female preference, exacted over generations, was the cause of exaggerated male traits, such as the tail of the peacock and, indeed, the tail-fins of swordtails. Darwin's idea is supported by a curious finding in some species of swordtail in which males are swordless: males with artificial swords, stuck on by impish experimenters, will suddenly find themselves more popular with females of the same species.

This idea led, naturally, to a scenario that described a trend in which species with long swords would evolve from a swordless ancestor, this evolution driven by a pre-existing female preference for ever longer swords. Phylogenies that included data on the tail would, naturally, presuppose that outcome. But the idea can really be tested only with data independent of the length or presence of the tail in males. Axel Meyer of the State University of New York at Stony Brook (now at the University of Konstanz, Germany) and his colleagues found those data in molecular sequence information. Using this information, they created a new cladogram for swordtails that showed how the sword is repeatedly gained and lost in evolution: and that the common ancestor most likely had a sword.[81]

This result throws the traditional scenario into confusion – yet although cladistics can be used to test such a scenario, one does not need any appeal to adaptation to justify the cladogram. Who can explain why swordtails gained and lost their swords so often? The result does raise an interesting conundrum about female preferences, for, rather than female preferences driving the evolution of a sword, the cladogram suggests that female fancy can persist in species that have *lost* their swords. Plainly, the whys and wherefores of swordtail armament are complex and transcend simple ideas about how it might have evolved – ideas conditioned by what we humans imagine possible, rather than by what really happened.

Apart from the kind of information used to make them up, cladograms built using molecular information are the same, conceptually, as cladograms built using, say, information about the pattern of cusps on teeth or the way bones are put together in a skeleton. This also applies to the concept of common ancestors, the hypothetical creatures that exist at the nodes of cladograms, if only as summaries of features. With parsimony as our guide, we can use present-day nucleotide sequences and work backwards, through the mutations, to the nodes, to get a good idea about the molecular sequences that existed in the common ancestry of a group of organisms that we are interested in. But with molecular sequences, we can go a step further – we can make those common ancestors come back to life. This much was achieved by a team of researchers in Switzerland, who reconstructed, synthesised and tested the digestive enzymes of the ancestors of ruminants extinct for tens of millions of years.

Ruminants – cattle, sheep, goats, antelope and deer – are among the most successful mammals in terms of numbers of species and individuals. It is easy to invent an adaptive scenario attributing the great success of the ruminants to their digestive systems, highly specialised for breaking down tough plant matter. Testing this scenario is more difficult.

One of the many powerful digestive enzymes that ruminants have at their command is pancreatic ribonuclease, secreted by the pancreas to digest a substance called ribonucleic acid, RNA (a close relative of DNA), in the bacteria that thrive in the ruminant gut. Like all enzymes, pancreatic ribonuclease is a protein, made according to instructions carried in a gene. Long ago in ruminant evolution, an ancestral gene for a general-purpose ribonuclease became duplicated and duplicated again. The copies diverged, each going its separate way, to produce different types of ribonuclease, found respectively in the brain, testes and pancreas. Freed from numerous other functions, the pancreatic form of ribonuclease could evolve without constraint into a digestive specialist. The ruminants were born and, with them, an evolutionary success story. But how can this story be tested?

The sequences of ribonuclease genes from ruminants and other animals can be used as the players in a cladistic analysis, to produce a cladogram – a picture of the evolution of ruminant digestion, from the point of view of the ribonuclease gene sequences in the animals concerned.

As with all cladograms, one can infer the palette of features that would have been present at the various nodes. We can infer, for example, that the common ancestor of myself and my cat Fred was a mammal with hair, mammary glands and so on. One can do the same with molecular sequence information. By comparing the sequences of pancreatic ribonuclease genes in any two ruminant species that have diverged from a common ancestor, it is possible to infer the nucleotide sequences of the pancreatic ribonuclease genes present in various common ancestors: such as, say, the common ancestor of antelopes and cattle, or the common ancestor of all ruminants, or even all even-toed ungulates (encompassing other animals besides ruminants, such as pigs).

Steven Benner and his colleagues at the ETH (Technical University) in Zürich, Switzerland produced a cladogram of ruminant pancreatic ribonuclease gene sequences, inferred the sequences that stood at the nodes, and used gene-sequencing technology to rebuild these gene sequences. Then, using the techniques of genetic engineering, the researchers introduced these artificial 'ancient' genes into *Escherichia coli* bacteria. The bacteria used instructions carried in the artificial genes to make pancreatic ribonuclease, the enzyme itself, in quantities that could be harvested and tested.[82]

The researchers found that 'ancient' ribonucleases – that is, the ribonucleases reconstructed from the sequences inferred from the cladogram, and then reproduced by genetically engineered bacteria – were biologically active substances, rather than useless goo. This in itself is a testment to the accuracy of the cladogram.

But enzymes have extremely specific modes of action. It is not enough to show that they are biologically active in some general way. A protease, for example, is an enzyme that will digest only proteins, spurning fats. A lipase, in contrast, will digest only fats; a methyltransferase will transfer only methyl groups rather than some other cluster of atoms, and so on. In the same way, ribonucleases are one-trick ponies. They digest ribonucleic acid. They happen to be extremely good at this, but that is *all* they do.

Surprisingly, ribonucleases from animals extinct for tens of millions of years, created by genetic engineering according to a cladistic recipe, worked in *precisely* the way you would expect a ribonuclease to function had it been extracted the day before from a fresh specimen.

The inferred common ancestor of all the ruminants digested RNA with the same efficiency as ribonuclease found in a modern cow. In contrast, more ancient ancestors – those found in the common ancestors of ruminants and other, non-ruminant ungulates – were more wayward, less tolerant to heat and much more active against RNA. These properties are seen today in the ribonuclease found in the testes of ruminants but not in the variety found in the stomach. These results support the idea that ruminant digestion was tied to a duplication in enzyme genes, perhaps more than forty million years ago. In this case, the results support the adaptive scenario in which the evolutionary success story of the ruminants is linked with their digestion. The armies of ruminants march on their stomachs, as well as their hooves.

Cladistics, then, can be used to help recreate the proteins and genes of long-extinct creatures. 'As sequences become available from a greater variety of species', wrote biochemist and molecular evolution specialist Caro-Beth Stewart in *Nature*, commenting on the Benner study, 'we should be able to reconstruct ancestral molecules going back further in time: perhaps – in the future – to the very root of the "tree of life"'.

In Michael Crichton's novel *Jurassic Park*, researchers recreate dinosaurs from DNA in blood cells, sucked up by mosquitoes which subsequently became entombed in amber and preserved as fossils. For the moment, this is science fiction; there is no absolutely certain, authenticated specimen of dinosaur DNA recovered from amber or anywhere else. Although DNA has been recovered from fossils dating back a few tens of thousands of years, most researchers agree that the preservation of intelligible DNA sequences in fossils millions of years old is unlikely in the extreme. All visits to Jurassic Park have been postponed indefinitely.

However, the work of the Zürich group suggests that even without DNA preserved as fossils, it should be possible to recreate any common-ancestral gene you please, provided that you have a good, even spread of modern sequences which can be used to make inferences about common ancestry.

Sadly, this rules out dinosaurs; the closest modern animals we have to dinosaurs are crocodiles and birds. This means that we might be able to recreate the genes of the latest common ancestor of crocodiles and birds, that is, the creature at the node whence crocodiles and birds diverged. However, this exercise would tell you little about dinosaurs in particular as it is likely that this creature was the most recent common ancestor of many creatures – not just dinosaurs, crocodiles and birds but also pterosaurs (pterodactyls) and other extinct reptiles. We would not be able to recreate, say, genes from *Triceratops* as distinct from those of *Tyrannosaurus rex*.

Nevertheless, one could disinter extinct genes to shed light on a host of interesting problems which, as in the ruminants, concern important lifestyle transitions. For example, molecular evolution suggests that whales are extremely close relatives of even-toed ungulates – the group that includes ruminants as well as pigs, camels and hippos – perhaps even forming a branch within the group. Some recent studies suggest that hippos are more closely related to whales than they are to other even-toed ungulates such as ruminants or other pigs.[83] These studies cast a whole new light on whales, forcing us to think of them as highly modified ungulates rather than representatives of a group of more obscure origins. Given that whales are descended from animals that once walked around on land, it would be fascinating to recreate genes from the common ancestry of whales and ungulates that have a bearing on the move from land back to water – genes concerned with respiration, digestion and other metabolic adjustments that whales made on their voyage back to the sea.

In the next few years we shall see an explosion in knowledge as researchers sequence the entire genomes – that is, the entire complement of genes – of an ever greater number of organisms. As yet, only a few genomes have been catalogued but the number will increase rapidly as technology makes it easier to sequence DNA automatically and in bulk. The current whole-genome catalogue includes a number of viruses and bacteria, such as *Escherichia coli*; some exotic microbes that live only in volcanic hot springs; a few disease agents such as *Helicobacter pylori* (the cause of stomach ulcers); the brewer's yeast *Saccharomyces cerevisiae*, and some others. A new bacterial genome is reported every few weeks. The most complex creature whose genome has been completely mapped is the roundworm, *Caenorhabditis elegans*, mentioned above. The genome of a plant – the thale cress (*Arabidopsis thaliana*) – will soon be completed, and likely targets in the near future include the laboratory fruit fly (*Drosophila melanogaster*) and the mouse (*Mus musculus*).

Within the next few years the genome of *Homo sapiens* should be ready for inspection. By then, the catalogue of organisms whose genomes are completely known should run into hundreds. The list will include the chimpanzee (*Pan troglodytes*), our closest living relative; our domestic animals; a huge range of disease agents; and curiosities such as the puffer-fish (*Fugu*) whose genome happens to be very small. There will be some notable omissions: the African clawed toad (*Xenopus laevis*), long the familiar of embryologists, will remain obdurate, because its genome is very large. The same is true, ironically, given their central place in the hearts of all true cladists, of the lungfishes – which, like amphibians, have very large genomes.

The application of cladistics to entire genomes will revolutionise our understanding of the history of life. For the first time in the history of biology, we will have a sure, testable method of reconstructing the branching pattern of the history of life, free from any preconceptions about function or adaptation. We will be able to apply it to whole genomes and gaze directly at the branches of the tree of life.

Cladistics is a way of looking at the world in terms of the pattern that evolution creates, rather than the process that creates the pattern. This means that it is free from assumptions based on particularities of cause and effect; as I mentioned above, a consequence of this is that cladistics can be used to examine any evolving system, free from any need to consider the peculiarities of the system. As we have seen, traditional evolutionary scenarios are forever tied to their substrates; for example, a scenario in which fishes acquire legs and walk on land cannot be applied to examine the evolution of birds from reptiles, because each scenario – each story – is unique. Cladistics, because it assumes much less about the evidence, reveals a great deal more.[84] It can be applied to any evolving system that produces, as a consequence of that evolution, a branching pattern of genealogy. The participants do not have to be living organisms for cladistics to work.

Palaeontologists have much in common with scholars interested in the history of medieval manuscripts. Before the invention of printing, the only way to reproduce a manuscript was to copy it by hand. The uniqueness of the original meant that most scribes only ever had access to a copy, rather than the original manuscript. Copies were made of copies which were then the fuel for further generations of copies, each one containing the accumulated idiosyncrasies of all the scribes and copyists who had worked on the text. Mishaps and misreadings built up in subsequent copies, just like genetic mutations – sometimes, whole lines of verse were missed, or new ones added.

Analysts of ancient texts are able to infer the history of copying – which manuscript was copied from which – by studying the accumulation of errors and producing a pattern, a kind of evolutionary family tree of errors, that assumes the fewest number of mistakes. This pattern is called a 'stemma'. As in a cladogram, all the participants are arranged along the top, except that in a stemma the participants are particular manuscript versions of a text, rather than fossils. From the pattern of errors, one can infer a parsimonious branching order and even guess at the existence and characteristics of missing 'common ancestors', manuscripts from which the extant readings were copied. For all practical purposes, a stemma is exactly the same as a cladogram. It obeys the same rules and conventions and yet the participants are not living creatures but books and manuscripts.

Adrian C. Barbrook from the University of Cambridge and his

colleagues applied cladistics to understanding the relationships of 58 extant fifteenth-century manuscripts of the Prologue to the *Wife of Bath's Tale*, part of Chaucer's Canterbury Tales.[85] The researchers isolated a group of little-studied manuscripts near the base – the root – of the cladogram. This basal position suggests that they diverged little from the common ancestor – Chaucer's original – and may in fact have been copied from it. Barbrook and colleagues urged Chaucer scholars to direct their gaze at this group of neglected texts. But the analysis suggested unexpected complexities – that Chaucer's own copy was not itself finished but a working draft containing the poet's own annotations, suggested alternative readings, notes on passages to be added or deleted, and so on. Far from the realm of fossil fishes, cladistics may have altered the course of medieval literary scholarship.

Linguists use very similar techniques when trying to trace the history of languages. In a pleasing union, linguists are increasingly teaming up with geneticists to trace how the evolution of languages tracks the movement of human groups in prehistory.[86] Because of the similarities in their methodological tool kits, linguists and geneticists have achieved an understanding so close that it is hard to tell where genetics ends and linguistics begins. Yet geneticists work with genes, which are products of organic evolution, and linguists work with languages, which are sets of symbolic representations of objects and abstractions – different things in substance and in concept, but whose histories are both accessible by the application of the principle of the sister-group relationship.

In Chapter 2, I despaired of ever being able to recognise an alien life-form for what it was. When we finally touch down on an alien planet, and encounter an entire new ecology like nothing found on Earth, how will we ever make sense of it, given that Earthly models are all we will have? How will we ever make sense of what we see, in terms of living organisms? The problem seems insoluble.

But the problem is really one of our own making because we are used to creating hypotheses of evolution that are tied to their individual contexts, each one tailored to a particular problem. A model of tetrapod evolution in which fossil fishes crawl ashore can only ever apply to tetrapods: it cannot be used to understand how other creatures, such as insects or plants, adopted a terrestrial existence. By extension, it cannot be used to help us understand how the nameless residents of the Jovian stratosphere, should they exist, came to do whatever it is that these creatures do. Traditional evolutionary systematics is tied to the Earth and can explain only Earthly things. If Earthly life is all we know, it is no surprise, then, that we will not be able to recognise an extraterrestrial as alive, even were it to stand right under our nose and ostentatiously wave its tentacles at us.

Cladistics knows no such restrictions. If cladistics can be applied to medieval manuscripts and ancient languages as to fossils and genes, it can also be applied to understanding alien ecologies, untangling the evolutionary family trees of extraterrestrials. It can do this by circumventing the problem of having to recognise aliens as living organisms, in the sense that Earthly organisms are living. But the state of creatures as alive or dead is a parochial conceit, which, for all we know, applies only to Earthly protoplasm. The example of the Martian meteorite, discussed in Chapter 2, shows that we cannot assume that what applies on Earth need apply anywhere else. By telling the history of life on Earth as a story, we look forever inwards, because the narrative method of traditional, process-based evolutionary biology is tailored to its terrestrial substrate. In the same way that the evolution of tetrapods tells you nothing about birds, the evolution of life on Earth tells you nothing about the evolution of any other system anywhere else in the Universe. The implications of this realisation are profound and shocking; they tell us that things we hold as fundamental, such as the principle of natural selection or even the very idea of 'life', may only be assumptions based on terrestrial prejudice. Cladistics, in contrast, makes no such assumptions. As a way of looking at the world, it is sufficiently general that the objects of its scrutiny are assumed to have changed only by some evolutionary process – the details of which do not matter – such that the results can be summarised as branching diagrams, be they the cladograms of biology or the stemmata of literary scholarship.

Cladistics has come a long way in half a century. Devised by an entomologist obscurely working behind the Iron Curtain and developed by palaeontologists working on fossil fishes, it is now routinely applied to solve problems in fields as diverse as medicine and medieval literature. Cladistics illuminates the history of the AIDS virus and will help us understand extraterrestrial life, should we ever find it. A view of life of such generality is sensitive to the inability of Deep Time to support narrative. For the same reason, cladistics may be our ally in the conquest of space, by helping us make sense of what lies beyond the Earth.

Researchers outside palaeontology now accept cladistics without a second thought as just another part of their tool kit. However, perhaps because cladistics went through its formative period among palaeontologists, it is among palaeontologists that the hearts-and-minds battles about cladistics are still being fought. Hennig, Halstead, Parrington, Patterson and Rosen are all dead. Gareth Nelson now lives mostly in Australia. Of the original Gang of Four, only Forey and Gardiner remain. The Cranley has gone – but it has not been forgotten: the quarrel between the Gang of Four and the palaeontological establishment has been taken up by a new generation of cladists. But the focus has moved, from fishes to birds – as I shall explore in the next chapter.

6 The Being and Becoming of Birds

Thou wast not born for death, immortal Bird!
No hungry generations tread thee down ...

John Keats, *Ode to a Nightingale*

Waterhouse Hawkins's dinosaur models in Crystal Palace Park are set among trees on an island in an ornamental lake. The visitor can view the dinosaurs only from across the water, the distance lending an air of timelessness to the concrete forms. The stately shapes of *Megalosaurus* and *Iguanodon* contrast abruptly with the squabbling of the waterfowl in the foreground. Even so, the dinosaurs are the focus of attention. The birds are such familiar parts of our everyday lives that we take them for granted.

Everyone instinctively knows what a bird *is*. Yet few stop to ask themselves the question of what, precisely, makes birds different from other creatures. What is the quality that unites ostriches and ospreys, setting them apart from basilisks and bandicoots? The question of what defines a bird – what constitutes this essence of 'birdness' – is at the centre of a debate of much current interest in evolutionary biology and palaeontology.

The debate touches on everything I have discussed so far in this book. It concerns the use of cladistics rather than adaptive scenarios for interpreting the history of life. It concerns the gulf between species defined as they are – as archetypes – or as evolving segments of a lineage; and, ultimately, between evolutionary stories told according to a comprehensible, human scale, and the qualitatively different vastness of Deep Time that admits no narrative structure.

So, then, what *is* a bird? Birds have skulls, and backbones made

169

of vertebrae, so they are vertebrates. Birds have four limbs – counting both legs and wings – so they are tetrapods. They lay eggs in which the chick is nurtured by a system of membranes including the amnion, so birds are amniotes. Birds form a coherent group of amniote tetrapods unified by several features not seen anywhere else in nature.

Birds have beaks instead of teeth. Their front limbs have become wings. Their breastbones are large, serving as anchors for powerful muscles; their collar bones are united to form a flexible spring brace which we call the 'wishbone'. Their backbones are tight, interlocked and stiff, the interlocking ribs contributing to a rigid cage. Yet many of the bones are hollow: lightweight yet strong, like tubular steel. Their tails are reduced to stubs. The pelvis and sacrum are welded together into a solid structure. Overall, the body of birds combines lightness with strength.

Birds have a remarkable breathing arrangement in which the lungs form just one part of a one-way air handling system that incorporates large air-spaces elsewhere in the body and even within the hollowed bones. Compared with reptiles and other tetrapods, many of the bones in the skull and hindlimbs seem heavily altered, fused together or lost. This suite of adaptations is not seen in any other tetrapod or vertebrate.

Most distinctive of all, birds have feathers. The fact that all birds alive today have feathers, unique structures not seen elsewhere in extant forms, has been taken as a useful marker, a 'key' feature. If an animal is a bird, then it must have feathers. Conversely, if it has feathers, then it is a bird, by definition. No animal without feathers is a bird.

It is the sum of these features that makes birds what they are. Together, they add up to that essence of birdness we seek. Now, this would be acceptable were birds expressions of an archetype, a separate divine creation, like Noah's Dove, forever distinct from Eden's Serpent. However, in our Darwinian world, we must find a way to accommodate the present state of birds – their *being* – with how they got that way – their *becoming*. So, as well as enumerating the list of things that makes birds what they are, distinct from anything else, we must also account for how the inventory of features that characterises birds came to be assembled in the familiar combination that allows us, without a second thought, to pair sparrows with sparrow-hawks to the exclusion of bats and bees.

This raises a conundrum. Birds are tetrapods, like you and me, *Acanthostega*, *Triceratops*, *Megatherium* and Fred the cat, so they must have evolved from ancestors which were, by definition, not birds. But if birds evolved from other creatures through a community of descent, steadily acquiring their bird-like features one by one, how and where can we draw the line between birds and non-birds? At what point – at which node in the cladogram – does an evolving bird collect enough bird-like features for us to recognise it as a bird, rather than as something else?

We know almost instinctively what it is that constitutes a bird at this moment in Deep Time, but what makes a bird in a wider, evolutionary sense? Given our present mental archetype of a bird – our search-image – it may not always be possible to spot an ancestor of birds in the fossil record, if that ancestor has only a few bird-like features, or none at all. If the presence of feathers, for example, is seen as an essential feature of birds, then we would not recognise a featherless fossil as a closer cousin of birds than of anything else.

However, should we find a fossil and convince ourselves that it is more closely related to birds than to other extant forms, we run the risk of tracing ancestry retrospectively: taking all the distinctive features we see in the present-day flock of birds – the wishbones, the beaks and feathers – and 'reading' them into the fossils, after the fact. As we have seen, fossils are mute, so it is tempting to tell their stories for them – stories told according to our own prejudices, through our own eyes, not through those of the fossil protagonists. These problems are all facets of the tension that arises when we adopt a picture of evolution that seeks to reconcile the dynamic of Darwinian evolution with the stasis of classical archetypes, yet takes no account of the qualitative difference between ordinary time and Deep Time. Such is the tension between *being* and *becoming* that must be resolved.

The traditional approach to the origin of birds sees their evolution as a matter of process, indissolubly linked with the origin of flight – their single, most distinctive habit. By blending the present adaptations of birds with their evolution, this scenario provides a simple, plausible and elegant solution to all the problems I have set out above. In the next few pages I shall discuss this approach in as convincing a way as I can. Later in this chapter, I shall show how and why it is wrong.[87]

The striking similarity of structure shared by all birds could reflect a response to a similar set of adaptive pressures. It is easy to see that flight is the principal source of these pressures. Almost all birds fly, and those that are flightless are *secondarily* flightless; that is, they show signs of having evolved from flying ancestors. As far as we know, no bird alive today is primarily flightless.

Heavier-than-air flight is an extremely energetic and dangerous activity. Natural selection will cull any flying animal not perfectly adapted to the task. This means that all birds will come to look rather similar, despite any differences in heritage. The features which make that essence of birdness are all connected with flight. It is this combination of features, all of them adaptations for flight, that makes this essence so distinctive. Feathers and the conversion of the forelimbs into wings are obvious adaptations for flight. Hollow bones and beaks (rather than teeth) are adaptations that save weight. Weight reduction is an obvious advantage for a heavier-than-air flyer and will be favoured by natural selection. The air sacs that fill the body also help to lower the overall density of the bird. The highly modified lungs allow for the processing of the large amount of oxygen required to power an activity as demanding as flight. The large, keeled breastbone anchors the voluminous muscles that move the wings. The wishbone and the rigid, box-like body provide a purchase for these muscles, bracing the body against their force, so that the bird does not crush itself with the power of its own wings.

Birds are not the only animals to have evolved flight. Although flight is demanding and difficult, it represents, in evolutionary terms, a passport to the air, which is a relatively unpopulated ecological niche. Insects, arguably the most successful animals both in terms of numbers of individuals and species, may owe their great success to flight. Flying insects are everywhere. Flightless insects, in contrast, are fewer. Among the vertebrates, and apart from birds themselves, the bats are fully adapted to powered flight, as were the extinct pterosaurs. Birds and bats are among the most successful vertebrates in terms of numbers of species and individuals. Because of flight, birds, bats and insects can colonise new habitats, such as distant lands and oceanic islands, that are accessible to terrestrial creatures only with difficulty. Flight is hard to accomplish, but the rewards are great and clearly repay the effort.

The problem of flight is not how to remain aloft – this much is ensured by a simple balance of forces – but how to get airborne at all. Perhaps the simplest method is to jump off a perch and break one's fall with a broad membrane that impedes the flow of air. This is the principle of the parachute. Thrust is provided free, by gravity, and all an animal needs to do is passively generate enough aerodynamic drag to ensure a soft landing.

Cats falling out of high buildings instinctively form themselves into parachutes. Once they reach terminal velocity, they relax and their limbs splay out horizontally. Cats that fall from intermediate heights – seven or eight stories – have not had a chance to adopt this posture and may be killed on impact. But there are reports of cats requiring only minor veterinary treatment after falls of more than thirty stories.[88] Cats, though, are only accidental parachutists. Several other animals are more accomplished, having evolved extensible, parachute-like membranes. The feet of the flying frog (*Rhacophorus*) are enormously expanded into windmill-like vanes; the body of the flying snake (*Chrysopelea*) flattens out as it leaps off a branch, forming a rudimentary but effective aerofoil, and members of a species of lizard (*Ptychozoon*) have flaps of skin on body and limbs that aid leaps from tree to tree. Parachuting is, therefore, a common habit, particularly among creatures living in dense forests. Animal parachutists leap from high branches, alighting lower down in the branches of adjacent trees. This form of locomotion, directly from tree to tree, is quicker and less energetic than climbing down one trunk, scampering across the intervening ground, and climbing up the next. Despite the risks of airborne mishap, it is also less dangerous as the parachutist is less likely to meet a predator in the air than on the ground.

Technically, parachuting refers to a mode of airborne travel in which the force of aerodynamic drag exceeds the generation of the opposing force, lift. Parachutists never gain altitude and their flight path is always more vertical than horizontal. A more efficient form of aerial locomotion is gliding. Like parachuting, gliding is a passive means of locomotion. However, gliders, unlike parachutists, may travel a considerable distance horizontally at the expense of only a modest vertical drop. This is achieved by the modification of membranes into aerofoils that generate lift as a consequence of thrust. An aerofoil moving horizontally is so shaped that the pressure of the air below the aerofoil is greater than that above, producing a net upward force.

Many animals are capable gliders. The flying dragon (*Draco*), a kind of lizard, glides, thanks to an extensive membrane stretched across extended ribs, and some of the earliest known fossil lizards had similar gliding membranes. Several mammals have flaps of loose skin extending from wrist to ankle which, when extended in flight on outstretched limbs, allow these animals to glide. This particular suite of adaptations is so effective that it has appeared several times, independently, in rodents such as flying squirrels; in the unrelated colugo or 'flying lemur' of South-East Asia; and in some marsupials.

None of these features, though, qualifies as powered, sustained flight. For this, the animal needs to do more than launch itself from a high place and let acceleration due to gravity do all the work. A true flyer must get airborne by exploiting favourable air currents; flapping to increase thrust and, through that, lift; or both. Because true flyers generate their own lift, they are no longer always constrained to take off from trees or cliffs. However, taking off from the ground requires a great deal of energy, as the animal needs to find a way of generating sufficient thrust before it can get airborne and, once it has started to take off, it must continue to maintain thrust and generate lift against gravity.

Of course, thousands of aircraft lift off every day in precisely these circumstances, and many birds take off from the ground or from water, so the process is obviously not impossible – yet aircraft and present-day birds are already accomplished flyers. The problem is imagining how birds evolved to get where they are today.

It is far simpler to imagine a proto-flyer with stubby little wings or rudimentary gliding membranes becoming airborne by falling downwards, exploiting gravity, rather than struggling upwards, in opposition to its pull.

Indeed, the present-day range of gliding and parachuting animals suggests how the evolution of powered flight in birds was achieved. The ancestors of birds could have been small, tree-living reptiles that adopted parachuting as a means of locomotion and avoiding predators. They did this by evolving extensive flaps of skin along the tail and the trailing edges of their forelimbs. Over time, the parachute-like structures became more extensive gliding membranes.

The advantages of the airborne life were great enough to prompt further elaborations. The animals could spend more time in the air, and travel further, if they flapped the forelimbs to generate thrust as well as lift, rather than simply extending the limbs in a passive pose. Scales became feathers, smoothing the contours of the body and reducing drag. The advantages conferred by powered flight on locomotion fed back into further elaborations: larger, more powerful flight muscles; a strong, box-like airframe; further weight-saving adaptations such as the beak instead of teeth, the fusion, hollowing-out or loss of bones and the elaboration of the system of air sacs. The enlarged lungs and the insulation offered by feathers became part of an improved metabolism, necessary to produce the energy required for active, powered flight. Over millions of years, tree-living reptiles evolved into birds.

When trying to reconcile the essence of birdness – the archetype of birds – with their evolution from some other category of organism, all one really needs to do is invoke flight as an adaptive stimulus. Without flight and the adaptations necessary for flight, birds would not be what they *are*; we would not recognise them as such. Virtually every feature of birds can be interpreted as an adaptation for flight and all birds have the same suite of adaptations. The evolution of flight is inextricably linked with the evolution of birds.

This is the traditional picture of the origin of birds. It is sensible and plausible, violates no established rules and invokes no unusual mechanisms. It is true to the Romer–Simpson view of evolution, in which long-term trends in the history of life may be explained in terms of adaptation and natural selection. For all that, the scenario that links birds with flight cannot be tested so your acceptance of it rather depends on whether or not you like it. As the person offering you this scenario, I can offer no greater guarantee of its truth than my own authority. In my command of the relevant facts, I must prove myself one of Colin Patterson's 'persons qualified to judge the evidence'.

However, my authority is no hedge against the possibility that other people, equally authoritative, might advance competing adaptive scenarios to explain the origin of birds. It is possible, for example, that birds evolved from animals that ran very quickly along the ground, flapping their arms, which might have originally been expanded into large vanes for a purpose other than flight – as display devices or as nets for catching insects. Palaeontologists have spent much time and effort weighing up the merits and demerits of each. But which is correct? Which is better? Ultimately, the choice is subjective, conditioned by how you feel about the strength of the evidence presented as well as the quality and authority of its presentation.

Given such an impasse, there is perhaps no other course than to examine the fossil evidence, to see if it can shed light on bird evolution. The first fossil bird was discovered in 1861, two years after Darwin's *Origin of Species* was published.[89] Richard Owen bought the crow-sized specimen, preserved on a slab, for his fledgling Natural History Museum, for the then huge sum of £450. He named it *Archaeopteryx* – 'ancient wing'. The specimen came from limestones of the Solnhofen district of Bavaria, deposited almost intact on the quiet floor of a tropical lagoon towards the end of the Jurassic Period, 150 million years ago. Since then, six more specimens of *Archaeopteryx* have come to light, all from these same limestones. Although robust in life, birds are extremely fragile in death; their hollow bones do not fossilise well and fossil birds are rare. *Archaeopteryx* owes its preservation to the exceptionally fine grain of the stone, quarried commercially for use in lithographic printing. The specimen of *Archaeopteryx* bought by Owen, now known as the 'London' specimen, is preserved in fine detail and includes impressions of the feathers; it was these that convinced people that *Archaeopteryx* was a bird.

Apart from the feathers, which look surprisingly modern, there is little that seems immediately bird-like about *Archaeopteryx*. It has jaws with teeth – there is no beak. The tail is long, bony and reptilian with a fringe of feathers down each side, rather than short and stubby, bearing a fan of feathers as is the style of modern birds. The wing finger-ends, reduced or absent in modern birds, bear functional claws. Some of the bones have holes that could have admitted extensions for the air-sac system,[90] but there are few other signs of weight reduction or a conspicuously high metabolism. *Archaeopteryx* could have flown far more efficiently than a sack of potatoes but may not have had the endurance or manoeuvrability of a modern bird.

During the late nineteenth and early twentieth centuries, remains of fossil birds were found from rocks dating as far back as the later part of the Cretaceous Period (around seventy million years ago). These birds were very much more like extant birds than *Archaeopteryx*. Until relatively recently, *Archaeopteryx* remained an anomaly, twice as old as the next-oldest fossil birds and extremely archaic in appearance. Nobody really knew where it fitted into the evolution of birds, except that, as it was by far the most ancient and most primitive bird, it fell rather easily into the role of ancestor or missing link.

And yet, for all the attention lavished on *Archaeopteryx* for more than a century (there is even an eponymous scholarly journal devoted to its study) and despite the illumination provided by a host of fossil birds discovered over the past decade or so, *Archaeopteryx* still remains enigmatic, something of a palaeontological Mona Lisa, an oracle, or a mirror. The messages people get from *Archaeopteryx* tend to reflect their own prejudices – their own search-images.

Ornithologists who study modern birds regard *Archaeopteryx* as an ancestor and an icon. Given that they have already judged where *Archaeopteryx* fits into the history of life, they look at the fossil and see exactly what they expect to find: bird-like features, such as the claws for perching, the well-shaped wings and, most important of all, the feathers, all but indistinguishable from the flight-feathers of a modern bird. *Archaeopteryx* has feathers, so it is a bird by definition. Its archaisms are only to be expected given the fossil's great antiquity when compared with other bird fossils. Because they study modern birds, ornithologists will naturally tend to see bird evolution in terms of perceived adaptations to birds' current, airborne niche. However, we know that there need be no connection between the present-day adaptations of a creature and the evolutionary origins of those adaptations. The conflation of the origin of birds with the origin of flight fails to account for that contradiction and that, ultimately, is why the traditional view of bird evolution – in which the origins of birds and flight go hand in hand – is wrong.

Palaeontologists, in contrast, come to *Archaeopteryx* with a different search-image. Unlike ornithologists, they see *Archaeopteryx* in terms more of its wider evolutionary relationships than of the perceived adaptive significance of features, such as feathers, seen in modern birds. To palaeontologists, *Archaeopteryx* looks very similar to members of a group of dinosaurs called theropods. These were bipedal creatures, usually interpreted as carnivores. Some theropods, such as *Tyrannosaurus*, were very large, but others were much smaller: the theropod *Compsognathus* was very small indeed, the size of a chicken. Fossils of *Compsognathus* have been recovered from the Solnhofen limestones of Bavaria, alongside *Archaeopteryx*. The anatomical similarities between *Compsognathus* and *Archaeopteryx* are so great that at least one specimen of *Archaeopteryx*, preserved without its feather impressions, has been mistaken for *Compsognathus*. In the 1970s, palaeontologist John H. Ostrom from the University of Yale (the mentor of Robert Bakker, who reconstructed *Triceratops* as a galloping animal) saw particularly close resemblances between *Archaeopteryx* and a subgroup of advanced theropods called dromaeosaurs, which included small-to-medium-sized dinosaurs such as *Velociraptor*. In this light, palaeontologists tend to see the presence of feathers in *Archaeopteryx* as intriguing decorations for the body of a theropod dinosaur rather than central, key features essential for explaining the course of evolution in birds.

Over the years, Ostrom and his colleagues accumulated a catalogue of anatomical similarities between dromaeosaurs and birds. These included hollow bones with holes for air sacs, three-toed limbs in which many of the usual reptilian complement of bones had been fused or lost and clavicles fused into wishbones.

The implications of these findings are profound and quite antithetical to the view that the origin of birds can be explained in terms of the origin of flight. Dromaeosaurs do not fit the picture of small animals that learned to fly by leaping out of trees. All those parachuting and gliding animals, advanced as examples of stages in the origin of flight, are small, quadrupedal climbers. Dromaeosaurs, in contrast, were bipeds, with long, spindly hind legs. The front legs were not modified into wings (in contrast, many dromaeosaurs had long-fingered, grasping hands) so the animals presumably could not have flown. Yet many of the other features they share with birds, such as hollow bones and fused clavicles, are conventionally regarded as adaptations for flight. More than this, the traditional view of bird origins depends on this adaptive interpretation being correct; if many bird-like features, usually thought of as adaptations specifically for flight, were found in non-flying animals, then the case for an intimate link between the origin of birds and the origin of flight is considerably weakened. Whatever the adaptive purpose for these features in dromaeosaurs, it was not flight.

The case of the dromaeosaurs parallels the story of *Acanthostega* and the origin of the tetrapod limb, as described in Chapter 2. The fact that *Acanthostega* had limbs with digits and yet was otherwise a fully aquatic animal – really, a fish with legs – was evidence against the usual scenario that the evolution of legs from fins was an adaptive response to the needs of locomotion on land.

Because the traditional view of the origin of birds depends absolutely on the central role of adaptation for flight, proponents of this view have gone to some lengths to discredit the idea that birds are related to bipedal dromaeosaurs rather than quadrupedal, tree-living ancestors. Here I discuss a selection of the objections that have been raised against a close bird–dinosaur relationship, showing how every one of these objections is flawed.

Rather than dinosaurs, enthusiasts of the traditional view cite a selection of small, fossil reptiles as pertinent to the ancestry of birds. An otherwise lizard-like fossil reptile called *Megalancosaurus*, for example, had a somewhat bird-like head; another, *Longisquama*, had a crest of elongated, scale-like structures on its back, reminiscent of feathers. The resemblances between these creatures and birds is tenuous but they fit the picture of small, tree-living ancestors better than do ground-living dromaeosaurs. They also lived at the right time – in the Triassic Period (248–213 million years ago) – to have been ancestral to *Archaeopteryx* and modern birds. Dromaeosaurs such as *Velociraptor* lived in the Late Cretaceous, long after *Archaeopteryx*, so they were in no position to have been the ancestors of birds.

However, the selection of Triassic reptiles such as *Megalancosaurus* and *Longisquama* as candidate bird ancestors, based on a small number of superficially bird-like features, is both selective and retrospective – an example of spotting trends in the fossil record after the fact. The search for the origin of birds should be informed by analyses of large suites of features, not a few here and there, marshalled selectively in support of a particular view.

The argument ruling out dromaeosaurs as close relatives of birds, on the basis of stratigraphic age, is similarly fallacious. It is true that *Archaeopteryx* lived 150 million years ago, and dromaeosaurs such as *Velociraptor* lived much later, around seventy million years ago. On this basis it seems odd, even outrageous, to claim that birds such as *Archaeopteryx* descended from dromaeosaurs. But the objection rests on the premise that one can take a series of fossils and string them together to create sequences of ancestry and descent. If that is possible, then, of course, Jurassic birds could not have descended from Cretaceous dromaeosaurs. However, we know that the creation of such sequences can only ever be made after the fact – they exist only in the mind of the beholder. But proponents of a close bird–dinosaur link make no such claims. Being cladists, they scrupulously avoid all notions of ancestry and descent created after the fact and concentrate on sister-group relationships. If you look back at Figure 12 in the last chapter and mentally label fossil 'a' as *Archaeopteryx* and fossil 'b' as *Velociraptor*, you can see that stratigraphic order is no barrier to any possible hypothesis of relationship. Given two fossils, or indeed two organisms, the most one can ever say is that they are cousins, united by a sister-group relationship. This is as much as can be justified given the evidence, and is independent of chronology.[91]

A more serious objection to the link between birds and dinosaurs concerns the anatomy of the hand in theropods and birds. The objection is serious because it concerns the anatomical evidence and is not based on flawed arguments about ancestry and descent or untestable adaptive scenarios about the habits of extinct animals. The hands of birds, although highly modified in the course of the evolution of wings, can be seen to have three fingers. The hands of dromaeosaurs also have three fingers. The question is whether bird fingers and dromaeosaur fingers are, in evolutionary terms, the 'same' three fingers. If they are not, the close link between birds and dinosaurs would be weakened.[92]

After the exuberant polydactyly of *Acanthostega*, *Ichthyostega* and other early tetrapods – whose limbs sported as many as eight digits apiece – tetrapods settled down to having a maximum of five digits per limb. Even these five could be lost in evolution. For example, the canonical picture of horse evolution starts with a five-toed ancestor that gradually lost its outer digits leaving the single, enlarged, third digit. A similar story can be told for theropods except that the digits that get lost are the little finger (digit V) and the ring-finger (digit IV), leaving just three digits: the thumb (digit I), the index finger (II) and the middle finger (III). The presence of five-fingered hands in early theropods, in which digits IV and V are clearly reduced, supports the theory that it is these two digits, not any other combination, that have been lost. In contrast, studies on the anatomies of modern birds suggest that birds have lost digits I and V – the thumb and the little finger – and so the three digits retained by birds are not I, II and III, but the middle three: II, III and IV. This suggests that the three-fingered state was achieved independently in birds and dinosaurs, the implication being that birds and dinosaurs are unrelated.

This objection to a close dinosaur–bird link, based on the evolutionary correspondence (or 'homology') of the digits, is derived from studies of how the hand bones form in chicks as they develop. Bones develop from cartilages, which in turn develop from knots of cells in the young embryo called 'cartilage condensations'. The argument rests on assertions about the future identity of particular knots of cartilage – whether a certain cartilage condensation seen in an embryo will, ultimately, become the bone of a certain digit in the adult. But cartilage condensations, like fossils, are never found with labels attached. A certain amount of interpretation is required before one can confidently assert that a particular knot of embryonic cartilage will always develop into a particular bone; such claims will always be open to interpretation. The situation is made more difficult by our obvious inability to study the embryology of dinosaurs so we will never be able to arrange a direct comparison.

By focusing on a single feature, such as the three-fingered hand, opponents of the bird–dinosaur view ignore the long list of similarities that have now been observed between birds and dromaeosaurs. However, when presented with this list, critics can always claim that similarities claimed as evidence of common ancestry were acquired independently in each case in response to similar adaptive pressures such as the need to support a high metabolic rate. This phenomenon – the separate and independent acquisition of similar features – is known as 'convergence'.

The claim of convergence is the most important objection to the bird–dinosaur link. Convergence is an extremely difficult problem. When confronted by two superficially similar organisms, it is not always easy to spot whether their similarity is a result of shared common ancestry or an adaptive response to similar pressures in otherwise unrelated animals. Ornithologists are very sensitive to this problem because many otherwise distantly related birds look deceptively similar because of a common response to the pressures of flight.

In a sense, claims of convergence are unanswerable, as one can always make a case, based on adaptation, that any feature thought indicative of common ancestry really represents a common response to similar adaptive pressures in unrelated species. Cladistics has a way of addressing the problem: the invocation of the principle of parsimony.

In Chapter 1, we saw how one cladogram – in which my cats, Marmite and Fred, formed a clade that excluded myself – was more parsimonious than another in which Fred and I formed a clade that excluded Marmite. The first solution is more parsimonious than the second because it requires the evolution of cat-like features (whiskers, pointed ears, retractile claws and so on) only once; the second solution requires the evolution of these same cat-like features on two separate occasions – in the branches leading, separately, to Fred and to Marmite. In the first solution, Marmite and Fred are similar by virtue of shared common ancestry; in the second, they are similar by convergence, the independent acquisition of similar features.

By convention, cladists choose, as a provisional hypothesis, the most parsimonious solution: the cladogram that requires the fewest evolutionary events to support its topology; in other words, the one that assumes the smallest amount of convergence. Of course, there is no law that says that evolution is always parsimonious. However, in a world in which it is very difficult and often impossible to decide whether similarity reflects common ancestry or convergence, it is pragmatic to adopt solutions in which convergence is minimal and start from there. Such solutions are no more than working hypotheses, subject to test, revision – even upset – in the light of subsequent evidence. But because they are true scientific hypotheses, cladograms make no claims to absolute truth in the Halsteadian sense. To claim convergence based on adaptive reasoning after the fact, however, is to claim privileged access to the kind of truth which is, in Donn Rosen's words, 'a truth that can be known', by setting yourself in the privileged position of an authority, a 'person qualified to judge the evidence'.

Birds and dromaeosaurs share so many of the same features that their independent, convergent origin seems unlikely. Cladograms in which these features are presented as having been acquired independently, by convergence, are highly unparsimonious. The most parsimonious cladograms, in contrast, are those in which birds and dromaeosaurs are closely related. These cladograms are, of course, still provisional hypotheses and it is possible that the similarities between birds and dinosaurs are convergent. If so, this will be revealed by cladistics in the light of new evidence but not by arguments based on untestable adaptive scenarios.

Why does the traditional, scenario-based view of bird origins fail? The reason, fundamentally, is because it starts with an untestable axiom, that flight has been crucial to the evolution of birds. The flight of modern birds is seen as the end result of a remorseless adaptive trajectory. Once this assumption is made, it is easy to make a story about how flight evolved by taking an assortment of unrelated animal gliders and parachutists and arranging them, after the fact, to justify the initial assumption – that flight has been the stimulus for the evolution of birds.

This is precisely the same line of argument that Gaskell used to justify the evolution of vertebrates from crustacean-like creatures. Gaskell assumed that Man represented the acme of creation. Because human beings are blessed with disproportionately large brains, Gaskell supposed that the evolution of vertebrates, culminating in Man, was driven by a progressive increase in the complexity of the nervous system. In the same way, it is easy to look at birds, pick flight as their most conspicuous achievement and use this as the basis for a retrospective tale of progressive aeronautical improvement to reach that goal.

As I discussed above, this argument is additionally flawed because it assumes that the current utility of a feature is a reliable guide to its history. Because gliding lizards, flying squirrels and so on all live in trees, proponents of the traditional view of bird origins assume that birds evolved from ancestors that were similarly arboreal. In support of this view, they look for arboreal adaptations in the extant fauna, such as the shapes of claws in tree-climbing birds, and claim to identify similar adaptations in fossils such as *Archaeopteryx*.[93] This is meant to stand in favour of the tree-living habit of the ancestors of birds and against the idea that birds could have evolved from bipedal runners such as dromaeosaurs.

However, we know that you can never make reliable statements about the habits of fossils based on modern models. Fossils are not living creatures and cannot be judged as such. They are the incompletely preserved fragments of creatures that once lived, not directly comparable with creatures alive today and which, as a consequence, can be comprehended only imperfectly in the light of modern models. *Triceratops* is like a rhinoceros only if you are prepared to imagine a ten-ton rhinoceros that laid eggs.

The argument based on the way organisms live now is spurious for the additional reason that adaptive purpose is hard to discern even in present-day animals. As I discussed in my critique of the 'aquatic ape' hypothesis, you could never guess, from their anatomies, that a wide range of mammals from voles to golden retrievers are capable swimmers. The same logic applies to claims regarding adaptation to life in trees. Goats, for example, are extremely good at climbing trees – surprisingly so, seemingly in defiance of their long, spindly, inflexible limbs, their lack of wings, grasping digits and prehensile tails – in short, their every feature that seems to adapt them for life on flat ground rather than up trees. By the same token, dromaeosaurs may look as if they were adapted to running along the ground but, if goats are any guide, dromaeosaurs might equally well have hopped up trees. In itself, the anatomy of dromaeosaurs provides no evidence against a tree-climbing habit, in the same way that curved claws, stereoscopic vision and even flight itself need be any evidence in favour of it.

The presumed habit of the supposed ancestors of birds – arboreal or terrestrial – is therefore quite irrelevant to retrieving the evolutionary history of birds from Deep Time. Presumptions about how the unicorn uses its horn are irrelevant to understanding the place occupied by the unicorn in the history of life and are, in any case, untestable. Cladists look at the anatomy of a creature without making prejudicial conclusions about what that anatomy *did* or what it might have been adapted *for*. Reason suggests no other course; once again, fossils are not living creatures but the petrified fragments of things which once lived in environments and ecosystems that are now lost. We must be careful how far we go in telling the fossils' stories for them. If the closest relatives of birds turn out to have been ground-living dinosaurs, then so be it – these animals presumably found some way of solving their aerodynamic problems, perhaps in a way that we cannot imagine using modern animals as models. And yet proponents of the traditional scenario go to some lengths in their efforts to discredit the bird–dinosaur link. They act as if any scheme not based on cladistic reasoning was worthy of scientific examination. Neither can ever be justified by advocacy; the only reason for supporting one or the other is through the independent validation of cladistics.

Furthermore, a scenario based entirely on one assumption – in this case, that the origins of flight and birds are closely coupled – is easily demolished if you can find fossils of non-flying creatures displaying features conventionally assumed to be adaptations for flight. Dromaeosaurs, which could not fly, have many such features such as fused clavicles, hollowed bones and air sacs. The very existence of dromaeosaurs exposes the weakness of the foundations on which rests the entire edifice of the traditional scenario of bird evolution.[94]

To many people, it is the feathers that make the bird. Without its halo of feathers, the London specimen of *Archaeopteryx* would have been seen as a dinosaur, not a bird. No matter how long the list of similarities, no matter how many times you insist that dromaeosaurs had very bird-like skeletons, there remains an emotional barrier to the full acceptance that birds and dinosaurs are close relatives. The fact is, birds have feathers while dromaeosaurs do not. Or, at least, they *did* not: the recent discovery of feathered dinosaurs has driven the final nail in the coffin of the traditional scenario.

In the early 1990s, the world of vertebrate palaeontology was buzzing with rumours of sensational fossil finds from China. Quarries in the province of Liaoning, to the north of Beijing, had yielded fossils of a primitive bird, *Confuciusornis*.[95] Like the first *Archaeopteryx*, the specimens of *Confuciusornis* were beautifully preserved, complete with feathers. Unlike *Archaeopteryx*, which had a conventional reptilian snout, *Confuciusornis* had a beak – the earliest record of a beak in the fossil record.

In more than 150 years, only seven specimens of *Archaeopteryx* have ever been found and each one is treated as a priceless relic. The contrast with *Confuciusornis* could hardly be greater; in only a few years, hundreds of specimens had been excavated from Liaoning province. *Confuciusornis* joined a steadily accumulating catalogue of fossil birds, unearthed in the 1980s and 1990s from a small number of fossil sites in China, Spain and other countries.[96] Most fossils came from the Mid to Late Cretaceous. None was as old as *Archaeopteryx* itself, which still remained the earliest known bird. *Confuciusornis*, though, came close; the age of the Liaoning fossil beds has been contentious, but they are now thought to be from the early Cretaceous, about 124 million years old,[97] about 25 million years younger than the late Jurassic Solnhofen limestones that have yielded *Archaeopteryx*.

But there were persistent reports that the Liaoning beds contained intriguing specimens of feathered dinosaurs as well as birds. The rumours came to a head in October 1996 at the annual meeting of the Society of Vertebrate Paleontology, held that year at the American Museum of Natural History in New York. In an alcove in a gallery devoted to the artefacts of the Pacific Northwest, surrounded by totem poles, Professor Chen Pei-Ji, from Nanjing, was showing snapshots of a feathered dinosaur to anyone who would ask, as casually as if they were pictures he'd taken on holiday. The pictures showed a slab from the Liaoning deposits. On the slab was a small dinosaur, preserved in side view, complete with impressions of soft tissues.

Most conspicuous was a set of brush-like fibres running in a crest along the back of the animal from head to tail, like a Mohican haircut. Patches of these fibres were to be found, less conspicuously, elsewhere on the specimen. Closer inspection revealed that these fibres branched in a manner reminiscent of the downy feathers of chicks. Sitting on a bench to one side of the hubbub was John Ostrom, by now the elder statesman of vertebrate palaeontology. He was amazed and pleased to see his suspicions about the close links between dinosaurs and birds come to such graphic fruition.

Later, when the fossil was formally described in the scientific literature, Chen and his colleagues claimed that these fibres might have had something to do with the origin of feathers. The belief was enshrined in the name he and his colleagues gave the fossil: *Sinosauropteryx*, the Chinese winged lizard.[98] At the time, there was a great deal of debate about the significance of the fibres. They did not really look much like either hairs or feathers. Chen and colleagues called them 'integumentary structures', as a way to avoid seeming to prejudge the functions or affinities of these structures. Some even supposed that they were not external at all but internal collagenous struts supporting a lizard-like frill.

The problem was that nobody had thought what a primitive feather might have looked like were we to stumble across one as a fossil. The feathers of *Archaeopteryx* – a fossil that is archaic in many other ways – look modern, each complete with a central stalk and marginal vanes. As such, they are hardly 'intermediate stages' in the evolution of feathers and offer no clues about the origin of these most distinctively bird-like features. As with so many unusual and interesting fossils, from *Acanthostega* to *Megatherium*, *Sinosauropteryx* represented an extinct form with no close modern counterpart, which we could comprehend only imperfectly through the distorting lens of modern forms.

The range of types of skin covering in extant tetrapods is limited; apart from bare skin, there are scales, hair or feathers, and that's it. The not-quite-feathery, not-quite-hairy fibres of *Sinosauropteryx* may represent a completely different, hitherto unknown variety of vertebrate skin covering, in the same way that the conodont elements discussed in Chapter 2 represent an extinct kind of tooth structure no longer seen in the modern world. In that sense, the significance of the fibres of *Sinosauropteryx* in understanding the origin of birds in particular is hard to estimate.

Cladistics, though, could offer a clue. Apart from the enigmatic skin covering, the skull and skeleton of *Sinosauropteryx* show that it is a theropod dinosaur. It is not a dromaeosaur but a representative of a more distant offshoot, branching from a very deep node in the theropod cladogram. This makes the position of *Sinosauropteryx* in the history of life extremely significant. It suggests that most if not all theropods could have had a tendency towards fibrous skin. If birds are close relatives of theropods, the implication is that the fibrous skin covering of *Sinosauropteryx* is indeed relevant to the origin of feathers.

On the other hand, these structures might just have been peculiar to *Sinosauropteryx*; after all, no known specimen of the dinosaur *Compsognathus*, a close relative of *Sinosauropteryx*, has any such structures even though specimens of *Compsognathus* are known from the same fine-grained limestones that have yielded *Archaeopteryx*, complete with its feathers. *Sinosauropteryx* remains an enigma: were its puzzling integumentary structures peculiar to itself, revealing nothing about the ancestry of feathers, or did they represent a significant discovery that might further understanding of the origin of feathers, and therefore of birds?

Answers could come only from more fossil finds. Further discoveries from China have produced – at the time of writing – two more dinosaurs clothed in *Sinosauropteryx*-like fibres. One is a dromaeosaur but the other is a therizinosaur, a member of an extremely aberrant theropod offshoot.[99] These results support the idea that many theropods, not just *Sinosauropteryx*, were covered in these fibre-like integumentary structures. Given that birds evolved from within the larger group of dromaeosaurs, the results also support the idea that these strange fibrous features really are intermediates in the origin of feathers – they show us what the first feathers looked like before they were 'feathers' in the sense that we understand the word, constrained as we are by what we see in modern birds.

After the announcement of *Sinosauropteryx*, the fossil rumour-mill intensified, with word that two other Chinese academics, Ji Qiang and Ji Shu-An, were working on two more small dinosaurs from the Liaoning beds. This time, the nature of the skin was quite unambiguous for these dinosaurs had unmistakable feathers rather than enigmatic fibres.

Ji and Ji had published a preliminary description of one of these dinosaurs, *Protarchaeopteryx*, in a Chinese scientific journal but had yet to publish a full account of either dinosaur in the West. To this end, they sought collaboration with two dinosaur experts from North America: Philip Currie, from the Royal Tyrrell Museum in Drumheller, Alberta, Canada, and Mark Norell, of the American Museum of Natural History. The fruits of this collaboration led to a paper on the dinosaurs *Protarchaeopteryx* and *Caudipteryx*, which was published in *Nature* in June 1998.[100]

Like *Sinosauropteryx*, these two dinosaurs are small, rangy and bipedal, with long, spindly legs and relatively short arms. *Protarchaeopteryx*, the larger of the two, is about the size of a turkey. It has a skin covering of fibres similar to those of *Sinosauropteryx* but with a switch of bird-like feathers at the tip of its tail. The other dinosaur described by Ji and Ji is called *Caudipteryx*. This is similar to *Protarchaeopteryx*, if a little smaller and with a relatively shorter tail. In addition to a fan of tail-feathers, *Caudipteryx* has a fringe of feathers along each of the trailing edges of its forearms like the go-faster fringes on a biker's leather jacket. The feathers are very like those of birds: each has a central stalk and vanes on either side.

Given the smallness of these creatures' arms, it is extremely unlikely that either dinosaur was capable of flight.

The preliminary cladistic analysis offered by Ji, Ji, Currie and Norell shows that *Protarchaeopteryx* is a dromaeosaur, a close relative of other small, bipedal theropods such as *Velociraptor*. *Caudipteryx*, however, is more bird-like: it forms a sister-group with *Archaeopteryx* and all other birds. It is therefore more closely related to birds than is either *Velociraptor* or *Protarchaeopteryx*. The researchers offered *Caudipteryx* as the closest known dinosaur relative of *Archaeopteryx*, perched on the very threshold of flight.

The implications of these discoveries are profound: the discovery of feathers in patently non-flying dromaeosaurs demonstrates that feathers existed before the evolution of flight. It can no longer be claimed that the origin of birds is inextricably linked with the origin of flight or denied that the heritage of the birds is closely linked with that of the theropod dinosaurs. Again, the cladogram showing that the closest non-flying relative of *Archaeopteryx* is a bipedal dinosaur, arguably a ground-dwelling runner, challenges the idea that birds evolved from small, tree-living reptiles.

The discovery of these feathered dinosaurs has brought the debate about the origin of birds to a close; the origin of birds must be sought among dromaeosaurs rather than some other group of dinosaurs or other reptiles. Birds are dinosaurs specialised for life aloft in the same way that tetrapods are lobe-finned fishes specialised for life ashore. The traditional view of bird origins, as well as being theoretically indefensible, has been trounced by the fossil evidence.[101]

Cladistics has played a central part in the bird–dinosaur debate as it did in the elucidation of the origin of tetrapods. To think of the pattern of life in terms of sister-groups rather than chains of ancestry and descent does away with any need to overstep the evidence and interpret it in terms of adaptation or purpose. So much seems clear from the way the debate about the origin of birds has gone within palaeontology. The media, however, still seem to adopt the copywriter's view of evolution as progressive, seeing *Sinosauropteryx*, *Protarchaeopteryx* and *Caudipteryx* in terms of missing links, stages in the evolution of birds. Although commentators have been forced to concede that feathers did not evolve for flight, they still seek causes. After Ji and colleagues published their paper in *Nature*, journalists would phone me for a quote. Their most urgent question was always about the purpose of the feathers for dinosaurs: if feathers did not evolve for flight, what *did* they evolve for?

As we now know, such questions about adaptive purpose are unanswerable. Fossils are not living organisms we can observe and on which we can perform experiments. They are the remains of creatures the likes of which no longer exist; creatures which lived in ecologies completely lost and which we cannot imagine except in the sketchiest terms; creatures which are interpreted imperfectly, after the fact, through the distorting glass of modern models which are always contingent and at best rough approximations. Any adaptive purpose we ascribe to structures in fossils may say more about our own prejudices than the structures themselves. So, although it is possible that the integumentary structures of *Sinosauropteryx* provided thermal insulation and that the feathers of *Protarchaeopteryx* and *Caudipteryx* could have been used for sexual display, these suggestions are based on modern analogues, on what we know (or suspect) about the structures and behaviours of modern birds. This course may not be valid because, as we know, the *only* feathered creatures in the modern world are birds so we are forced to interpret any feathered fossil in an avian light using current utility to speculate about evolutionary history.

As we have seen, current utility and evolutionary history are two different things that need not be connected. To confuse the two is to confuse *being* with *becoming*, ordinary time with Deep Time, to produce a voodoo palaeontology in which dinosaurs are viewed retrospectively as stages in an evolutionary trajectory that would *necessarily* culminate in the appearance of birds; such was a trap into which the traditional interpretation fell. The very existence of bizarre creatures such as feathered dinosaurs, unfamiliar to the modern world, should be clue enough that these creatures could have employed their feathers in ways in which we cannot imagine, ways which – like the creatures themselves – have no close correspondent in the modern world. As I wrote about the unicorn, it is futile to speculate about the function of its horn unless we have some knowledge of the unicorn's place in nature.

Cladistics allows the creation and testing of hypotheses about the relationships of creatures, independent of any concerns about adaptation or function. Speculations about the adaptive purpose of the feathers of dinosaurs are untestable and thus unscientific. As I have shown, it is sometimes possible to test hypotheses about adaptation but the subjects should be observable as living creatures. Hypotheses without tests are no more than cocktail-party chatter and are without value except perhaps as entertainment – they are not science.

Fossil discoveries made over the past few years have, in addition, challenged the retrospective view of bird evolution as linear and progressive, starting with a reptilian ancestor of some kind and progressing inexorably to birds. Several kinds of theropod dinosaurs are now known to have had fibrous body coverings similar to that of *Sinosauropteryx*; it now seems that feather-like integument was not uncommon among theropod dinosaurs of various kinds. The arm-bones of *Unenlagia*, a theropod found in Argentina, suggest that the animal could have tucked its arms into its body in the same way that birds furl their wings. The dinosaur *Velociraptor* had a wishbone and hollowed bones in the manner of birds. There is a fossil of the dinosaur *Oviraptor*, petrified while seated on its nest, incubating a clutch of eggs and using its forelimbs to shield the nest from the sandstorm that finally buried mother, eggs, nest and all – in a pose much like that of a farmyard chicken.[102]

Evidence for bird-like anatomy and behaviour in theropods now seems uncontestable. But this interpretation carries the danger of using a search-image derived from modern birds and interpreting the features of extinct creatures (in this case dinosaurs) in an avian light. Of course, such over-interpretation is possible but this is the risk we inevitably run when trying to make sense of fossils in terms of living animals. The problems of making sense of what you see are great enough without making life harder by requiring the fossils to conform to an additional layer of prejudice about adaptive purpose. What the current crop of bird-like dinosaurs definitely shows is that the origin of flight has no particular connection with the origin of birds. Feathers did not make the bird because feathers existed long before anything that we would recognise, instinctively, as a bird. Theropods were dragons indeed but ones that took tens of millions of years to get airborne. Crucially, bird-like features can be seen in a variety of theropod offshoots, from *Sinosauropteryx* to dromaeosaurs, but not in a single lineage – a fact that weakens the view of bird ancestry as progressive, linear and driven by a single, overriding adaptive necessity.

The emerging consciousness of the origin of birds among dinosaurs has allowed us to see *Archaeopteryx* in a more critical light. Once upon a time, *Archaeopteryx* stood alone as the earliest fossil bird. Its uniqueness made it an icon, conferring on it the status of an ancestor. But this status is ill-deserved: the existence of *Sinosauropteryx*, *Protarchaeopteryx*, *Caudipteryx* and an increasing number of others shows that *Archaeopteryx* is just another dinosaur with feathers. The fact that *Archaeopteryx* might have been able to fly is not an issue if – as we have seen – there is no logical reason why the origin of birds; should be linked with that of flight. This realisation knocks another hole in the conventional wisdom about the evolution of birds; there is a whole literature that explores the possible flying capabilities of *Archaeopteryx*, the rationale being an exploration of the origin and evolution of flight in *birds*. But if *Archaeopteryx* is just another feathered dinosaur, such work loses its impact – you may as well seek to understand flight in birds by studying flight in unrelated flying animals such as bats or bees.

The removal of *Archaeopteryx* from its status as the canonical earliest bird raises questions about what, precisely, we mean by the term 'bird'. At the beginning of this chapter, I said that everyone knows what a bird is, almost instinctively, but that this intuition need say nothing about how birds evolved. To repeat the questions asked earlier: where can we draw the line between birds and non-birds? At what point – at which node in the cladogram – does an evolving bird collect enough bird-like features for us to recognise it as a bird rather than as something else? If *Archaeopteryx* cannot be assumed, automatically, to have been a 'bird' in the sense in which we intuitively grasp the term, what does the term 'bird' really mean? Which animals should be included in the term 'bird', and which should not?

There are three solutions to this problem, all of them unsatisfactory. The first is to confine the term to those members of a clade comprising the most recent common ancestor of all living birds. This clade would exclude *Archaeopteryx* and would also exclude many creatures such as *Confuciusornis*, with feathers, beaks and the ability to fly. Were we to meet a live specimen of *Confuciusornis*, we would immediately think of it as a bird. However, it seems unnecessarily restrictive to exclude *Confuciusornis* and other extinct species from birdhood on what would amount to technicalities of anatomical structure.

On the other hand, we could reserve the term 'bird' for any creature more closely related to modern birds than to the next most closely related living sister-group – the crocodilians, in this case – irrespective of its habits or anatomy. This solution goes to the other extreme because many extinct creatures, including dinosaurs and pterosaurs, which are more closely related to birds than to crocodilians, would be classified as birds. To imagine pterosaurs as birds is just about credible but the admission of creatures such as *Triceratops* or *Tyrannosaurus* to the ranks of birds suggests that the use of the term in this sense is so inclusive as to be meaningless.

The third solution is to reserve the term 'bird' for those members of a clade comprising the most recent common ancestor of *Archaeopteryx* and all living birds. This would include *Confuciusornis* but exclude dinosaurs and pterosaurs. This seems like the answer to our problem, but it is not – it is a pragmatic and entirely arbitrary sop to what we humans, who have grown up in a world of feathered, flying birds, are happy to think of as 'birds'. Considered objectively, it is very difficult to justify placing the dividing line between birds and non-birds at any particular node.

The problem is that the category 'bird' relies for its justification on the use of grades as well as clades in classification. This is another instance of the argument about the salmon, the lungfish and the cow discussed in the previous chapter. Everybody thinks they know what a fish is but the term 'fish' has no strict scientific meaning: it simply refers to our mental picture of what we expect a non-tetrapod vertebrate to look like. In the same way, although we instinctively know what a bird 'is', we are appealing to a mental archetype of what, in our experience, birds 'should' be, not to some scientifically validated list of characteristics.

The dethronement of *Archaeopteryx* has a further, even more surprising implication. If *Archaeopteryx* is just another feathered dinosaur, it becomes possible that there were non-flying dinosaurs even more closely related than *Archaeopteryx* to modern birds. Such remarkable dinosaurs – or, perhaps, flightless birds – have been known for more than a decade.

Since the 1920s, the Gobi Desert of Mongolia has yielded a rich treasure trove of fossils, mostly dating from the later part of the Cretaceous Period, the heyday of the dinosaurs. Many dinosaur skeletons have been found there, of course, as well as dinosaur eggs and nests, skeletons of tiny mammals still in their burrows, and many other creatures of all kinds.[103] In 1987, a Soviet–Mongolian expedition brought home a partial skeleton of a curious turkey-sized dinosaur. Five years later, in 1992, the ongoing joint expedition of the Mongolian Academy of Sciences and the American Museum of Natural History found another animal of the same kind. There was enough to see what kind of animal these bones belonged to and offer a formal, taxonomic description with a name.[104]

This creature, named *Mononykus*, is very thin and rangy, with a tiny head on a serpentine neck and extremely long, spindly legs, each terminating in a bird-like foot, with three long toes in front and a 'reversed' big toe. The fibula – the smaller, more splint-like of the two shin bones, the other being the tibia – is small and does not extend all the way down from the knee to the ankle. Seven of the vertebrae are fused to the pelvis to make a structure called a 'synsacrum', and the hip socket is braced by a distinctive bony flange, the 'antitrochanter'. All these things are distinctive features of birds, but not of dinosaurs of any kind – not even dromaeosaurs, or even *Archaeopteryx*.

The most unusual features of *Mononykus* are its forelimbs which are as short and robust as the hindlimbs are long and spindly. Although each forelimb is no more than about ten centimetres long, it has a thick humerus, a robust elbow joint for the attachment of strong muscles, and a big wrist bone – a fusion of several smaller bones into a single blocky unit. This fusion is a characteristic feature of the reduction and simplification of bones in the bird forelimb during the evolution of the wing. The reduction of digits is another such feature and in *Mononykus* it has proceeded to an almost grotesque degree; all digits have been lost except one, but this remnant is massively expanded and ends with a huge claw. The breastbone, lying on the chest between these two strange limbs, was robust with a small bony 'keel', presumably for the attachment of the muscles corresponding with the flight muscles of modern birds.

Mononykus is a strange creature but not unique. In the past few years, the fossilised remains of similar animals have been discovered in Patagonia and North America. One of the South American ones, *Alvarezsaurus*, was originally thought to be a small dinosaur but is now regarded as a relative of *Mononykus*. *Mononykus* and its relatives are now grouped together as the Alvarezsauridae.

The latest addition to the family is a fossil called *Shuuvuia*, which, like *Mononykus*, comes from Mongolia.[105] This fossil is especially valuable because it has a complete skull, the first known for any alvarezsaurid. It has a number of features otherwise seen only in birds. For example, the orbit (eye socket) is confluent with various spaces behind it, opening up the entire cheek region. In reptiles, the rear edge of the orbit is marked by a bar of bone that projects down from the postorbital bone, immediately behind the eye, until it meets the jugal bone at the back of the upper jaw near the jaw hinge. In birds, this downward spur fails to meet the jugal. This feature is as diagnostic of modern birds as the presence of feathers – and it is also seen in the alvarezsaurid dinosaur *Shuuvuia*.

Alvarezsaurids could not have flown. The arms, so small and stubby, could not have been wings; whatever these strange structures were adapted for (and speculation is futile), it was not flight. However, many features of the skeletons and skulls of alvarezsaurids suggest that they might have been closer cousins of modern birds than *Archaeopteryx* – even though *Archaeopteryx* has feathers and wings. This hypothesis seems extraordinary. However, its strangeness comes not from the facts but our own preconceptions, which tell us that creatures with wings and feathers, such as *Archaeopteryx*, must be related to birds. But this preconception comes from the false logic that links current utility with past history.

To mix a metaphor, flight was the albatross that laid the golden eggs. On the one hand, it gave birds the key to a new ecological niche – they became masters of the air. On the other, flight is so difficult and energetic an exercise that birds give it up as soon as the opportunity presents itself. The world is full of flightless birds. The extinct dodo of Mauritius is only the most famous example of birds whose ancestors, having landed on remote islands free from ground-dwelling predators, lost the ability to fly. The penguins represent an order of birds whose entire membership is flightless: the ratites (ostriches and their relatives) represent another. These groups are both extremely ancient, perhaps going back more than fifty million years. The fossil record contains early evidence of the abandonment of flight. *Hesperornis*, for example, was a flightless – indeed, almost completely wingless – seabird that lived in the Cretaceous Period, seventy million years ago.

If the alvarezsaurids were closer cousins of extant birds than *Archaeopteryx*, it is possible that they – like penguins, ostriches and *Hesperornis* – were secondarily flightless. But superficially, alvarezsaurids look more like small dinosaurs than birds. This raises a host of interesting questions such as whether other flightless dinosaurs did not have flight in their ancestry; the antiquity of flight among birds and dinosaurs; whether some of the feathered dinosaurs so far recovered had flying ancestors; and indeed, if many familiar theropods, such as *Tyrannosaurus* or *Velociraptor*, had flying ancestors. These bipedal, carnivorous theropods could have been the flying dragons that fell to Earth.

We do not know whether any of these wonderful and terrifying possibilities are true. The truth, as we have seen, is not the point – the point is that we are able to ask the questions at all. The conventional view of bird evolution starts from a preconceived scenario in which birds and flight go together and birds evolved from small lizard-like reptiles that lived in trees. A mind schooled in this cramped, stultifying and intellectually impotent scenario simply cannot ask such questions because it cannot admit any evidence that challenges the prior assumption; if it did, the entire scenario would collapse.

How blinkered this view now seems, when we can, with justification, populate the ancient Earth with real dragons, whose plodding effigies in Crystal Palace Park are now surrounded by a babble of downy dragonets, dabbling and clucking after scraps of bread thrown to them by small children! Yet this view is no more blinkered than the tales we tell about our own species, which forms the subject of our concluding chapter.

7 Are We Not Men?

> The dark hut, these grotesque dim figures, just flecked here and there by
> a glimmer of light, and all of them swaying in unison and chanting:
>   'Not to go on all-Fours: *that* is the Law. Are we not Men?'
>
> H. G. Wells, *The Island of Dr Moreau*

It is 14 February 1993 – Valentine's Day – and I am in San Francisco
on an assignment for *Nature*. It's Sunday so I award myself the day
off and drive my rented Mercury to the University of California,
Berkeley, to visit a palaeontologist called Tim White.

For the past few years, White and his colleagues have been
studying the very earliest part of the hominid fossil record, discovered
in the Ethiopian desert. Less well-known than his work in Africa is his
definitive study of cannibalism among the Native Americans of the
American Southwest. Proving, from mute fragments, that people ate
one another is very hard. Every scratch is minutely examined for signs
of deliberate intent. White shows me his collection of archaeological
material, smashed-up mealtime scraps from human midden heaps.

Almost as an afterthought, he shows me a set of monochrome
contact prints, the first photos of the fruits of his team's latest field
season in the Afar region of Ethiopia. Each picture shows a bone
fragment that looks less like a fossil than a cornflake. The finds are
4.4 million years old and come from a place called Aramis. 'This is
the earliest-known hominid', says White, proudly, but with a touch
of self-deprecating humour that demonstrates a sensitivity to the
inevitably piecemeal nature of human fossil remains, in which all the
evidence for the hominid lineage between about ten and five million
years ago, several thousand generations of living creatures, can be
fitted into a small box.

We look at the photos again. I am trying to make sense of them. It might have helped to have seen the originals but the Aramis collection is in Addis Ababa awaiting preparation. 'You'll get a manuscript about it in a year and a half', says White. This sounds leisurely but preparing fossils – extracting them from the rock, cleaning them, conserving them, studying and describing them – takes a long time, especially when the fossils concerned are half a world away.

A year passes. In the late spring of 1994, I return from lunch to the *Nature* office to find a yellow sticker on my phone. 'Mr Asfaw called', reads the note, giving a number. Berhane Asfaw, a palaeontologist and a close colleague of Tim White, has accumulated an enviable list of fossil discoveries. Like the Kenyans I was to meet four years later, he has an eye for finding the one meaningful pebble in the whole jumbled desert. I dial the number, reaching a small hotel in South Kensington, one of dozens of modest establishments in the shadow of the Natural History Museum.

Half an hour later I'm in the hotel lobby. Berhane Asfaw emerges from the elevator carrying a large brown envelope. Most people send their manuscripts to *Nature* in the mail – a few, concerned with secrecy and mishap – carry theirs by hand, even across continents. Asfaw is in transit between Berkeley and Addis Ababa, he explains: London is a convenient stopover. We go to the basement bar of the nearby Norfolk Hotel for a suitable toast to the Aramis Papers. For there are, in fact, two manuscripts. In one, by White, Asfaw and their colleague Gen Suwa of the University of Tokyo, the crushed cornflakes whose pictures I saw more than a year ago and 5,000 miles away have metamorphosed into the base of a skull, some teeth and an assortment of bone fragments. Some of the teeth are set in fragments of jaw similar to the fossil that Robert will pick up at LO5 in the western Turkana more than four years later. The fossils from Aramis form the basis of a new species, *Australopithecus ramidus* – later renamed *Ardipithecus ramidus* – the earliest known hominid.[106]

Every feature of *Ardipithecus ramidus* seems primitive[107] compared with other hominids, yet the shape of the teeth suggests kinship with later creatures and the shape of the base of the skull suggests that the spinal column entered the skull from underneath as in a biped, such as a modern human, rather than from behind as in a quadruped, such as a chimpanzee. Were *Ardipithecus* any less human-like, it would be hard to decide whether it were a closer relative of chimpanzees than humans.

The second manuscript, by a larger team comprised of Giday WoldeGabriel of the Los Alamos National Laboratory and his colleagues, concerns the environment in which *Ardipithecus ramidus* once lived. Fossils of seeds, monkeys and woodland antelopes recovered at Aramis indicate a woodland habitat. If so, then we were seeing for the first time a hominid from that part of the story before hominids left the forests.

Nature published the two manuscripts on 22 September 1994. The news made the front pages of newspapers in London, New York and around the world. 'The bone that rewrites the history of man', thundered Nigel Hawkes on the front page of *The Times* in London, above a picture of the tooth-bearing jaw fragment. Tim White got his picture in *Newsweek*. The incomparable British tabloids gave the story upbeat coverage. 'Yabba Dabba Doo!' cheered Ashley Walton in the *Daily Express*, turning *Ardipithecus ramidus* into Uncle Ram, Fred Flintstone's honoured ancestor.

'We have unearthed man's missing link: he's not much to look at. But more than four million years ago he could have been the boy next door. Uncle Ram, whose remains have just been discovered in a remote desert, is believed to be man's earliest ancestor. He may be stunted, hairy, almost toothless and not exactly handsome, but this chimp-style creature is our long-lost relation'.

The day before, Bill Mouland of the *Daily Mail* had called me for a quote and I had described the significance of the finds to him in lively language. About halfway through the piece in Thursday's paper stood a paragraph in ambiguous isolation: 'Dr Gee believes he may end up holding his place as the oldest human ancestor'. I remember all this so clearly because the next day was my wedding day. The *Daily Mail* quote found its way into the Best Man's speech and I thanked my in-laws for allowing me to become the 'missing link' in their family.

In all the press attention, *Ardipithecus ramidus* had been tagged as a 'missing link'. *Nature* had encouraged this because, at the time, it seemed a more digestible substitute for phrases such as 'the-hominid-closest-to-the-evolutionary-split-between-our-lineage-and-that-of-the-apes' which, while more accurate, were also clumsy. The justification was the extremely primitive nature of *Ardipithecus ramidus*. In a commentary in *Nature* accompanying the Aramis papers, Bernard Wood of the University of Liverpool (now at George Washington University in Washington DC) wrote that it was a testament to the maturity of paleoanthropology that *Ardipithecus ramidus* did not overturn the current consensus but quietly slotted into the family tree in the place reserved for just such a creature. If *Ardipithecus ramidus* hadn't been discovered, we would have had to invent it.[108]

This line of thinking carries vestiges of the old, classical view of the system of nature in which each species occupies its own discrete, preordained place in the scheme of life. As I described in Chapter 4, Darwinism did not supplant this view; instead, the idea of change was grafted on to the older picture, inserting arrows of change between adjacent archetypes in the Great Chain of Being. As editors of *Nature* we were, on reflection, wrong to pander to the voodoo palaeontology as portrayed by the media because it presupposes a model of evolution that is linear, upwards and progressive. We know that this model is mistaken and yet it is deceptively easy to see evolution in this way, especially when we are discussing our own origins. After all, we are the descendants that get to write the book so we should be particularly careful not to expect the sparse hominid fossil record to bear narrative scenarios based on the steady, linear acquisition of the adaptively useful traits that make us human.

Despite decades of patient work in pitiless places, we still know comparatively little about the evolution of humanity. What follows is a cartoon version of what happened, according to the latest evidence. I have told it as plainly as I can, to show how hard it is to justify any kind of narrative based on the evidence.

Comparisons of the genes of humans and chimpanzees suggest that the lineages leading to modern humans and chimps diverged around five million years ago. Few hominid fossils are known from the interval between five and ten million years ago and all of these are uninformative scraps. There are no known fossils, of any age, that might illuminate the ancestries of the extant African apes.

The interval between five and three million years ago saw the appearance in the fossil record of several different hominid species. *Ardipithecus ramidus* is the earliest known in any detail. Another, *Australopithecus anamensis*, discovered by Meave Leakey and her team at Turkana, was a primitive species that lived in the area around four million years ago. Fossils of a third species, *Australopithecus afarensis*, have been recovered in the Rift Valley from deposits dated between about 3.6 and three million years ago. A fourth species, *Australopithecus africanus*, is recorded from cave deposits in southern Africa of a slightly younger age. All these creatures were bipeds. However, the interrelationships of these creatures are not beyond dispute and it is possible that further species of hominid remain to be discovered from this interval.

The East African environment appears to have changed between three and two million years ago. The climate became drier, and the forests were replaced by more open country. Many species of mammal, relatively common as fossils in sediments older than about 2.5 million years, are rare or absent from younger sediments. At the same time, many new species appear. Palaeontologist Elizabeth Vrba has linked this 'pulse' of faunal change with changes in climate. Documenting wholesale changes in fauna demands a systematic and comprehensive approach to fossil-collecting.[109] By collecting fossils from around the 2.5-million-year mark, Louise Leakey and her team hope to gather enough data to test Vrba's 'turnover pulse' idea. However, recording a change in the inventory of wildlife is a different matter from finding the causes of that change, as we saw in Chapter 4 in connection with the extinction of the dinosaurs.

Several different hominids appear in the fossil record between three and two million years ago. Some of them fall into a well-defined group distinguished by large, heavily muscled jaws and pavement-like teeth. *Paranthropus boisei* (the 'Zinj' of the Leakeys) from Tanzania, *Paranthropus robustus* from southern Africa and *Paranthropus aethiopicus* (the 'Black Skull') from the western Turkana represent this set of hominids. Other hominids recovered as fossils from this interval are less easy to categorise. In their skeletons they are perhaps more like *Australopithecus africanus* than anything else although they tend to have larger brains. The names of these species – *Homo rudolfensis* and *Homo habilis* – show that they are conventionally regarded as members of our own genus although some researchers such as Bernard Wood have questioned this, suggesting that they may be no closer to modern humans than are some other extinct hominids.[110]

The earliest known hominid that looked more like a human being than *Australopithecus* was a creature known as *Homo ergaster*, which appeared in Africa some time before two million years ago. *Homo ergaster* was very similar to another species, *Homo erectus*, which appeared in the fossil record about two million years ago. It is arguably the earliest hominid known outside Africa.[111] Its remains have been found throughout tropical and temperate Eurasia, from Spain in the west to China and Indonesia in the east.

The period after around 800,000 years ago is marked by a variety of human-like creatures from across the Old World. These are conventionally referred to as *Homo heidelbergensis* after a jawbone discovered in Germany near the city of Heidelberg. Hominids from Spain dated to around 300,000 years ago resemble those of Neandertal Man (*Homo neanderthalensis*), the classic Ice-Age European whose bones were first discovered in a cave near Düsseldorf, Germany, in 1856. This creature lived in Europe and western Asia and survived in the mountains of Spain until as recently as 33,000 years ago. Other, less well-known regional forms appeared in eastern Asia.

The first fossils that can be referred to our own species – *Homo sapiens* – appeared in Africa some time before 200,000 years ago, although a recent discovery of a skull in the Danakil Depression of Eritrea might put this back as far as a million years.[112] *Homo sapiens* starts to appear outside Africa from about 90,000 years ago, first in the Middle East and then in South-East Asia and Australia (60–50,000), Europe (40,000) and the Americas (13,000 years ago).

Some researchers think that *Homo sapiens* evolved in several places at once, from local populations of earlier species of *Homo*. For example, modern Europeans could have evolved from

Neandertals. Other researchers think that all modern humans descend from a relatively small population of more recent, fully modern migrants from Africa. Most evidence at present favours the latter view.[113] The distinction between modern humans and Neandertals has been reinforced by a study by Svante Pääbo of the University of Munich and colleagues who extracted DNA from Neandertal bones and compared it with DNA from modern humans. It turns out that Neandertal DNA is not only very different from that of modern Europeans but completely outside the known range of variation in *Homo sapiens* DNA.[114]

The work on Neandertal DNA picked up on a line of research which, in 1987, gave a boost to the idea that all modern humans owe their ancestry to fully modern humans that migrated from Africa. The late Allan C. Wilson of the University of California, Berkeley and his colleagues collected DNA from people of many racial groups and origins, compared the similarities and differences between their sequences and used the principle of parsimony to arrange these sequences into a cladogram.[115]

The deepest branching point – the most basal node – in the cladogram presented by Wilson and colleagues made a clear distinction between people of exclusively African origin and people from Africa and everywhere else. This suggests that the richest and most ancient reservoir of human diversity is to be found in Africa. Humans outside Africa constitute a more or less homogeneous offshoot of one particular African stock. Based on this result, Wilson and colleagues speculated that the most recent common ancestor of all modern humanity lived in Africa around 200,000 years ago. This work has since been subject to a barrage of technical criticism[116] but has been revised and extended and supported by other work. The case for a recent African origin for modern humans remains very strong.

The clarity of this work owed something to the type of DNA studied. Most of the DNA that constitutes the genetic material resides in the nuclei of cells, bound up in bodies called chromosomes. The variation of living organisms is a function of sex, the merging of the nuclei of egg and sperm and the shuffling of genes so that offspring have the attributes of both parents.

But there is more to genetic material than chromosomes in the nucleus. Small bodies called mitochondria, located outside the nuclei of cells, have their own independent DNA arranged in a single 'chromosome'. Mitochondria are the power stations of the cell, responsible for cellular 'respiration' (technically, the release of energy from food). Having DNA on site allows mitochondria to produce a few of the essential enzymes it needs without having to import genetic instructions from the nucleus.

This mitochondrial DNA is small when compared with nuclear DNA but it has an advantage attractive to researchers studying evolutionary divergence: mitochondrial genes are not shuffled and reshuffled with each generation. Except in extremely unusual circumstances mitochondrial DNA is passed down, intact, through the maternal line. This is a consequence of the mechanics of fertilisation. An egg cell is enormous – the largest individual cell in the human body, with the possible exception of a few mature nerve cells. The egg has to be large as it needs to carry sufficient fuel to nourish it through several cell divisions after fertilisation. This process requires energy so egg cells are well stocked with mito-chondria, each of which contains its own DNA. A sperm, on the other hand, is possibly the smallest cell in the human body. It is divided into a head – little more than a DNA-packed nucleus – and a lashing tail, driven by a battery of mitochondria. On fertilisation, the sperm enters the cell, leaving the tail behind: any interloping sperm mitochondria degenerate after entry.

The genetic material of the resulting embryo is consequently biased. The genetic material in the nucleus – the bulk of the DNA in the cell – is of paternal and maternal origin, in equal parts. But *all* the mitochondrial DNA derives from the mitochondria in the egg alone. The sperm contributes nuclear but not mitochondrial DNA. So, although most of your genes come from your father and mother, subsequently shuffled together into a unique combination, your mitochondrial DNA, in every cell of your body, comes exclusively from your mother and is not mixed with other genetic material. Your mother got her mitochondrial DNA exclusively from your grandmother with no contribution from your grandfather. And *she* got it from *her* mother, and so on. This is true of every human that ever existed.

The only way that mitochondrial DNA can change is by the slow accumulation of mutations. Mutations in mitochondrial DNA (or, indeed, nuclear DNA) convey information about relatedness in much the same way as any anatomical feature. For example, it is more parsimonious to say that two mitochondrial DNA sequences that share a certain mutation are more closely related to each other than either one is to a third mitochondrial DNA sequence that does not carry that mutation.

By comparing the mitochondrial DNA from a range of people and plotting the degree to which mutation had changed it through time, Wilson and colleagues came up with a cladogram that reflected, strictly, the pattern of maternal lineage among modern humans. The pattern suggested that mitochondrial DNA of all modern humans has its origin in a single variety of mitochondrial DNA present in people in Africa around 200,000 years ago – the age inferred from assumptions about the mutation rate in mitochondrial DNA. The fact that mitochondrial DNA is inherited maternally gives the work a particular cultural and emotional resonance because, carried to its logical conclusion, the work implies that everyone gets their mitochondrial DNA from a single individual, the many-times-great-grandmother of us all. Wilson and his colleagues spelled this out very clearly in their report: 'All these mitochondrial DNAs stem from one woman who is postulated to have lived about 200,000 years ago, in Africa.'

It is possible to get carried away and rewrite this sentence with 'Eden' for 'Africa', and 'Eve' for 'one woman'. It would be easy to read too much into this; in particular, we could attach too much importance to the personification of Eve herself. In Africa, 200,000 years ago, she was not the only woman but one of many, each with her own complement of mitochondrial DNA. Eve could not have known that chance would make her, alone, the sole ancestor of modern humanity. Her sisters, friends and cousins could have been as fecund as she was, perhaps more: except that their particular lineages died out.

We do not know for certain that even this last statement – that every 200,000-year-old lineage became extinct, save one – is true. All we can say is that the more we look, the picture of a single African origin looks stronger, not weaker. However, only a few hundred of the several billion humans now living have been sampled so there remains the possibility that there are people, somewhere in the world, with varieties of mitochondrial DNA that fall outside the entire modern human range sampled so far. If these people live outside Africa, the hypothesis of a single African origin will be weakened.

There is, in fact, evidence for such aberrant mitochondrial DNA. The Neandertal DNA extracted by Pääbo and colleagues is mitochondrial, not nuclear. A cladistic analysis shows it to be an outgroup to the whole of modern humanity. In this way, cladistics has tested – and refuted – the polygenetic hypothesis that modern Europeans, as opposed to modern humans from elsewhere in the world, have a Neandertal ancestry. If this hypothesis were true, the cladogram would have placed Neandertal mitochondrial DNA alongside modern European DNA, to the exclusion of the DNA from modern humans from elsewhere in the world.

The story of mitochondrial Eve should be tempered by our knowledge of the limitations of the evidence and the behaviour of Deep Time. That is, whereas it must be true that all modern humans share a common ancestry, we cannot know with certainty who that ancestor was or precisely where she lived. Although it is possible to go to Africa and unearth the remains of a human being that lived 200,000 years ago (and such remains have indeed been found), you would have no way of knowing for certain whether this individual really was your direct ancestor. As we know, fossils are isolated points in Deep Time that cannot be connected with any other to form a narrative of ancestry and descent.

By concentrating on relatedness rather than ancestry, cladistics can be used to sketch the outlines of human evolution. Figure 13 shows a simplified cladogram of hominid interrelationships. It shows how *Paranthropus* and *Homo* are separate offshoots from *Australopithecus*. This suggests that the genus *Australopithecus* – like the class 'fishes' – does not constitute a 'proper' group. Members of the genus *Australopithecus* form a grade rather than a clade, unified by their general primitiveness rather than by any distinguishing features. *Ardipithecus ramidus* is an outrider as befits its primitive status. The cladogram makes no presumptions about who is ancestral to whom, for such things cannot be known for certain. There are no 'missing links', no chain of ancestry and descent, no sign of progressive advancement towards the acme that is humanity.

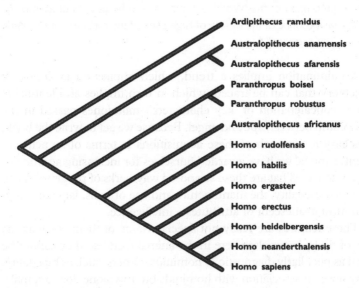

Figure 14. A cladogram of hominid interrelationships. Diagram by Majo Xeridat[117]

If the view of evolution as linear and progressive is ingrained into the popular imagination, evolutionary biologists probably have only themselves to blame. In his textbook, *The Life of Vertebrates*, the late J. Z. Young – a prominent educator of evolutionary biology in Britain – set out vertebrate evolution explicitly in terms of progressive improvement. Mammals and birds represent more organised entities than, say, reptiles, amphibians and fishes because they display greater complexities of structure which require an ever more refined degree of physiological control. Given that humans are mammals, it is easy to see them as a culmination of this trend. He writes:

We shall expect to find in the mammals even more devices for correcting the possible effects of external change than are found in other groups. Besides means for regulating such features as those mentioned above we shall find that the receptors are especially sensitive and the motor mechanisms able to produce remarkable adjustments of the environment to suit the organism, *culminating in man* with his astonishing perception of the 'World' around him and his powers of altering the whole fabric of the surface of large parts of the earth to suit his needs. [My emphasis][118]

A culmination implies a trend, which implies an evolutionary trajectory that can be traced, which in turn implies an identifiable linear concatenation of 'key characters'; milestones passed in the long climb between ape and angel. Because we get to write the book, it is easy to see each of these acquisitions in terms of an achievement, a medal we have awarded ourselves for increasing aptitude or competence. What are these supposed waymarks of progress? They include bipedality, the manufacture and use of tools, language, and that most evanescent of attributes – intelligence.

These features are really not special. Some of them, such as the use of tools, occur elsewhere in the animal world so they cannot be used as peculiarly diagnostic of hominids. Others, such as bipedality, may have an association with hominids but this alone does not make these features particularly unusual as other animals may display equally interesting yet analogous features. As regards intelligence and language, we must ask ourselves whether their perceived importance is a distortion because it is *we* who are the subjects of our own investigation. After all, we are trying our best to be objective about deeply held prejudices for the unique place that humanity generally believes itself to occupy in nature. We are not looking dispassionately at a boxful of beetles.

The problem with our perception of human evolution as linear and progressive – 'culminating in man' – results from our perceived zoological uniqueness; all our close relatives are extinct. Uncle Ram, Lucy, Zinj, Neandertal Man, Mitochondrial Eve – they are all gone, banished to Deep Time. Because the residents of Deep Time are unable to tell us their stories, we have to invent their stories on their behalf – stories coloured inevitably by our own self-aggrandising perspective. If only some of these ancients were still with us. If they were, they would be able to offer us their own perspectives on the meaning of life and existence and, perhaps, force us to see ourselves as others would see us.

A reader of an early draft of this book reminded me that we are not as unique as we might like to think, at least, not in terms of genetics. Chimpanzee and human DNA differs by about one per cent – a separation far less than that which distinguishes, say, species of fruit fly. But the degree to which we differ from other animals, in terms of numbers of nucleotides or some other quantitative measure, is not my argument; the problem lies in the *perception* of this difference. Because we are humans and have no closer relatives than dumb chimpanzees that are still extant and who might offer some badly needed perspective on our state, we like to exaggerate the differences between humans and other animals to make ourselves seem special, a breed apart. This is why the smallness of the difference between human and chimpanzee DNA always seems so surprising.

Bipedality – the habit of walking on our hind legs – is regarded as a diagnostic, hominid feature. Because we are bipeds, we expect this to be a general characteristic of all hominids, in the same way that the possession of feathers is conventionally seen as a defining characteristic of birds. But to link bipedality with hominids is to make the same mistake that traditional zoology makes with feathers and birds; it confuses current utility with past history, *being* with *becoming*. Looking at the apes and hominids as a whole, bipedality is just one of a number of unusual forms of locomotion. Gibbons, for example, use their arms to swing from tree to tree, a style of locomotion called 'brachiation'. The gorilla and chimpanzee are quadrupeds but walk on their knuckles rather than on their palms or soles, like bears (quadrupeds of comparable size), or the tips of their digits, like cats or ungulates. Brachiating, knuckle-walking and bipedality are found in different members of a close-knit group but each is, in its own way, a distinct and specialised form of locomotion. There is therefore nothing special, advanced or progressive about bipedality; only the fact that it is *we* who are bipedal, and it is *we* who are writing the book, makes it so.

Along with the acquisition of bipedality, increasing brain size is generally seen as something that tracks human evolution. Increasing brain size is usually related to increasing intelligence. It is certainly true that mammals have bigger brains relative to body size than fishes, amphibians or reptiles but this is a rather crude generalisation. The trend disappears when one looks at a set of different yet closely related species such as living and fossil hominids. To complicate matters, brain volume can vary enormously among individuals in a species with no discernible connection to intelligence.

Humans today have a brain volume of around 1,300 cubic centimetres, but this varies widely (the writer Anatole France was notoriously small-brained, which goes to show that brain size and cleverness are not correlated). You, therefore, cannot use brain volume as a simple benchmark for separating humans from non-humans. Yet prescriptions for the membership of our own genus *Homo* (as distinct from genera of extinct hominids such as *Paranthropus* or *Australopithecus*) have repeatedly included a brain volume of greater than a certain amount, such as 1,000 cubic centimetres.

Such a definition fails to accommodate the degree of variation one sees in modern humans. Next time you are on a crowded bus or train, especially in a large cosmopolitan city such as London or New York, take a good look at yourself and the other passengers. The range of shapes and sizes in modern humans is quite incredible. The height in adult humans ranges from less than four feet to more than eight – a factor of two, in a single species.

Given this degree of variation, how can we know that the small sample of fossils that have been left to us tells us anything useful about the range of form accessible to an extinct species? It is possible that a large male *Paranthropus boisei* could have had a brain larger than a small adult female *Homo ergaster*. But which would have been the more intelligent?

Despite these problems, palaeoanthropologists might try to use a quantitative measure such as brain volume[119] and use this as a proxy for the relative development of intelligence. However, the relative intelligence of members of the same species – *Homo sapiens* – is notoriously difficult to measure and there is much debate about the significance of various measures of intelligence or even whether such measures are meaningful at all. Presumably, even greater caution should be exercised when seeking to compare the intelligences of *different* species, such as, say, *Homo sapiens* and *Homo erectus* – even more so when one of them is extinct. Clearly, before we can get any grasp on the evolution of intelligence, we need to ask whether intelligence can survive as a distinct, unitary quality that can be compared across species as well as between individuals within a single species.

I do not think that the concept of intelligence can survive in any meaningful way when comparisons are sought between different species. The reason is that intelligence is surely no single thing but the name we give to certain ways in which organisms interact with their environments: ways which seem sympathetic to human motives and actions (for it is we, after all, who are watching and measuring). But different species see the world through different eyes and those eyes need not be human ones. The adaptations and needs of one species differ from those of another. It may not be valid, therefore, to compare the intelligence of Flipper the dolphin with that of my cat, Fred. Flipper and Fred belong to different species and their brains, senses and anatomies are adapted to serve different ends in different environments; presumably, Flipper and Fred have entirely different ways of looking at the world. So if Fred and Flipper are both intelligent, they are somehow intelligent in different ways that may not be comparable. If so, then this must apply when we seek to compare human intelligence with the intelligence of other hominids such as Neandertal Man or *Homo erectus*. These were different species from us so presumably they saw the world in different ways.[120] Comparisons between fossil species suffer from the additional problem of total inaccessibility of experimentation. We cannot ask extinct creatures to run through mazes for us.

There is, however, a deeper problem with this quixotic act of estimating intelligence across species. I alluded to this above: how can we be sure that we are interpreting the 'intelligent' behaviour we are seeing in an objective way, or do we mark as intelligent any behaviour which is recognisable in terms of our own? Do dogs and cats seem more intelligent than aardvarks and anteaters because they interact with us in sociable yet subservient ways which we find flattering? Do we rank intelligence according to our perception of how much or how little a species behaves 'like us'? If so, it is not surprising that the view of the evolution of intelligence has tended to be rigorously linear, a trajectory leading inexorably to the acme of humanity.

The problem is that we, as a species, are so alone. There are no species with whom we can converse meaningfully about our different ways of looking at the world. The chimpanzee, our closest extant relative, is yet not close enough for a decent conversation. Everything closer is long extinct. Because we are alone, we tend to think that our point of view is the only one possible. On the other hand, meaningful exchange between different species might be undermined or even made impossible simply *because* our views would be different.

Another attribute regarded as distinctively human is the ability to manufacture and use tools. The earliest stone tools that we can distinguish from randomly chipped cobbles appear in the fossil record around 2.5 million years ago. When the Leakeys found the remains of the hominid *Paranthropus boisei* at Olduvai Gorge, they also found tools. It has always been assumed that the tools were made by *Homo habilis*: his name means 'handy man', in recognition of this inference of tool use. However, it is possible that the elevation of *Homo habilis* to near-humanity has been a retrospective award by ourselves based on an association between fossils and stone tools that could be mistaken. After all, stone tools are never labelled with their maker's mark. It is possible that the tools were made by *Paranthropus*, or some other unknown author, and early *Homo* never made tools at all.[121]

Even if we knew for certain that *Homo habilis* made the tools credited to him, the tool-making habit is no guarantee of humanity. Many kinds of animals make and use tools as sophisticated as the earliest flint artefacts. A species of crow living on the island of New Caledonia in the Pacific modifies the vanes of leaves in various ways to make picks and saws for wheedling insects out of crevices, but this crow is not a hominid.[122]

Once one extends the definition of tool to any artificial structure external to the body of the animal one must admit the nests of weaver birds, the honeycombs of bees, the mounds of termites and even coral reefs and stromatolites – reef formations created by bacteria – to the roll-call of non-hominid manufacturing ability. Apologists for human intelligence might say that these examples of manufacture are the products of instinct (birds, ants, termites) or physiology (corals, stromatolites). What sets human manufacture apart is a quality known as 'planning depth'. Human beings create tools with a purpose. They can 'see', in their mind's eye, what the tool is going to be used for. But even if this point is moot in the case of the New Caledonia crows (after all, you cannot see into the mind of a crow to gauge its intentions), it is also debatable for many cases of human manufacture.

Perhaps the most distinctive artefact from the entire history of humanity is the Acheulean hand axe. It gets its name from the site of St-Acheul in France where it was first described, but these hand axes have been found all over the Old World from Britain to Indonesia and were made in more or less the same way for a million years. They are conventionally seen as part of the tool kit of the species *Homo ergaster* and its close relative *Homo erectus*, the first hominid definitely known to have occurred outside Africa.

Hand axes are so well known that people can reconstruct, in detail, how they were made. But still, nobody knows what they were *for*. And you may well ask whether a tool whose model did not change for a million years really is a tool in the sense that a modern human would understand the word. If the pattern of an implement lives and dies unchanged, with a particular species, over the course of ten thousand centuries, you cannot assume that it is any more the product of intelligence than the nest of a weaver bird, brilliantly – if blindly – contrived. *Homo ergaster* was perhaps the first species whose body proportions looked human but the same humanity need not have applied to its mind. After all, *Homo ergaster* belonged to a different species from *Homo sapiens* and presumably had a different outlook on the world. In *The Wisdom of Bones*, Alan Walker and Pat Shipman suggest that *Homo ergaster* was a savannah predator like a hyena or a lion, neither of which needs to be able to think grand thoughts in order to survive.

Ideas about intelligence tend to confound our appreciation of the significance of that other human attribute – language. It seems nothing separates humans and animals so obviously as language. What other species can convey its thoughts so precisely and in such detail? But we are the only species we know that uses language, so we have no standard for comparison. As a consequence, we tend to play down the richness and subtlety of visual, auditory and olfactory communication found among organisms right down to bacteria, and focus on the subtlety of human language as if it had everything to do with intelligence. Language permeates everything we do to such an extent that we can hardly avoid making the link. How often have you imbued some inanimate object with human capacity, just because it addresses you in human language? Humour finds a rich seam in the way people react to things such as puppets, speaking clocks, recorded messages, computers, parrots and so on, according them a humanity they do not deserve by virtue of their intelligence – simply because they can speak.

This conceit is exposed by two notable explorations of the human condition; the fourth part of *Gulliver's Travels* by Jonathan Swift, published in 1726, and *The Island of Dr Moreau* by H. G. Wells, which appeared in 1896, almost two centuries later. One can hardly imagine two authors with more divergent views of the world: Swift, the dystopian, pessimistic cleric and Wells, the utopian, optimistic atheist. Yet their novels have many parallels – and some telling inversions.

In his fourth voyage (the previous three being narratives of the hero's exploits elsewhere), Gulliver finds himself washed up in a country inhabited by the Houyhnhnms, a race of intelligent horses. They seem to be intelligent because they have a spoken language (which Gulliver learns to speak) and prove to be civilised and decorous hosts, if rather cold and rational.

Life for the Houyhnhnms would be perfect but for the fact that their country is infested with bands of animals called Yahoos. These creatures revel in squalor and strife. They are completely barbaric and appear to have no more language than evil grunts and gestures. Gulliver shares his horror of Yahoos with the Houyhnhnms and this is deepened by the shock that, but for clothes and language, he too would be classed as a Yahoo. This revelation is cemented when he is asked by his Houyhnhnm hosts to describe the conditions in his own country as honestly as possible (for the Houyhnhnms are incapable of lying and tend to take things literally).

This is the central part of the story, in which Swift – through Gulliver – delivers a satire on the squalor and corruption of Britain in the early eighteenth century. A powdered wig and fine words are all that separate us from being Yahoos.

The literal-minded Houyhnhnms take this line of reasoning to its extreme. For all his good manners and his mastery of the Houyhnhnm language, Gulliver is still a Yahoo and is ordered to leave their country. Once back in England, Gulliver finds that he cannot bear the company of his fellow human beings. In the end, he feels just about able to endure their existence provided they do not start to find reason for optimism in their wretched condition. Given that we are all Yahoos, Gulliver condemns such error as vainglorious:

'My reconcilement to the Yahoo-kind in general might not be so difficult, if they would be content with those vices and follies only, which nature hath entitled them to . . . but when I behold a lump of deformity and diseases, both in body and mind, smitten with *pride*, it immediately breaks all the measures of my patience; neither shall I be ever able to comprehend how such an animal and such a vice could tally together.'

The Island of Dr Moreau starts in a similar way. Prendick, a young scientific man from London, is cast away on a desert island whose inhabitants include Moreau – a notorious vivisectionist hounded out of London – his drunken assistant, Montgomery, and a small population of what Prendick takes to be intensely ugly natives. Although each is deformed in its own way, the natives are capable of speech in a charming semi-archaic fashion as if straight out of William Blake. At first Prendick thinks that the natives are human beings, the disfigured results of Moreau's dreadful experiments. He starts to wonder whether, as an uninvited guest on the island, he will end up similarly mutilated.

In a long speech, Moreau explains that these 'Beast People' are not vivisected humans but animals such as pigs and bears, raised by vivisection to a near-human state. This, says Moreau, is easier to do than one might think:

'Very much indeed of what we call moral education is such an artificial modification and perversion of instinct; pugnacity is trained into courageous self-sacrifice, and suppressed sexuality into religious emotion. And the great difference between man and monkey is in the larynx . . . in the incapacity to frame delicately different sound-symbols by which thought could be sustained.'

The Beast People are animals, and the possession of spoken language counts more for their humanity than their origins; it was enough to mislead Prendick into thinking that they were humans. In their grotesque chanting of the 'Law', language allows the Beast People to express a kind of religious belief, conditioned as it might be by their fear of Moreau. The story ends as Moreau and Montgomery are killed by their creations and Prendick is marooned on the island with a band of Beast People who are losing their powers of speech and reverting to their original animal natures. Once rescued and back in England, Prendick – like Gulliver – cannot bear the presence of his fellow humans, and for the same reasons, if strangely inverted:

'I could not persuade myself that the men and women I met were not also another, still passably human, Beast People, animals half-wrought into the outward image of human souls; and that they would presently begin to revert, to show first this bestial mark and then that.'

The parallels between *Gulliver's Travels* and *The Island of Dr Moreau* have not been lost on the critics. In *Trillion Year Spree*, their critical history of science fiction,[123] Brian Aldiss and David Wingrove mark the Beast People as 'kinfolk' to the Yahoos, as expressions of humanity's essentially animal nature. But this fails to recognise the inversion. The Yahoos are humans that do not speak and are initially dismissed as animals; the Beast People, in contrast, are animals that can speak and so are initially regarded as human. The line between human and animal is vanishingly thin and is demarcated by language. To the reader, the horror of both novels rests in the shock of being told how thin this line is when, as humans, we tend to think that the possession of language counts for a great deal and is in some way a mark of human intelligence.

Language, then, seems to be loaded with more notions of human intelligence than it should have. It is, though, distinctly human. In *The Language Instinct*, cognitive psychologist Stephen Pinker revives and explains Noam Chomsky's idea of a 'universal grammar'.[124] That is, the basic rules for language seem to be hard-wired into our brains; they are not learned, and so require no intelligence to master. Children the world over acquire language quite naturally at similar stages and can construct complex sentences without having to be taught. All that is required is that they grow up in a milieu that uses language. It seems that all languages are based on the same underlying principles of grammar and no society has ever been found in which language does not occur. If this scheme is true – that learning to talk is as easy as learning to walk – then there can be no correlation between language *per se* and intelligence. Even the least articulate human being of otherwise normal mental capacity is vastly more articulate than the most refined chimpanzee.

None of this should be seen as any belittlement of the seismic effect that language has had on our species. Language, through the permanence of writing, has permitted human beings to store a great deal of information outside the brain. Once writing had been invented, people no longer had to rely on memory. They could build on the achievements of generations rather than starting everything from the beginning each time. This has created in the past 10,000 years (and especially the past 500) a dramatic acculturation leading to an interdependence of human societies and their parts and, through that, the creation of entirely new epiphenomena – things that happen by virtue of the complexity of human society but which cannot easily be controlled by it. I refer to such things as stock-market movements and economic cycles, commonly discussed as if they were entities in their own right rather than the unpredictable results of myriad smaller, yet interacting, human processes.[125] Nevertheless, the whole of civilisation, the acquisition of culture, the drive ever onwards and upwards and outwards builds, at root, on the ability of human beings to share information by means of language, a product not of intelligence but, like breathing, of the normal physiological function of the human body.

So much, then, for the special humanity inherent in all those steps to greatness: bipedality, toolmaking, intelligence, language. Is there any way we can regain our self-respect? A critical look at all these achievements shows that they are based more on the things we *do* than the things we *are*. It is hard to pinpoint the origins of humanity in the fossil record if many of its essential features depend on things that do not fossilise, such as motives, thoughts and the ability to use language.

One way to get round this is to come up with a prescription for humanity that relies strictly on the hard evidence, to define *Homo sapiens* just as we would any other species, on the basis of features that all its members share and which are found in no other species; in other words, to rest our notion of humanity on the dry and technical-sounding list of molars, mandibles, skulls and bones on which taxonomy depends.

But is this not somehow unsatisfying – especially if it is *we* who are being described? Surely, no list of bones and teeth can tell us what being human really *means*? Are we not more than subjects prey to the unfathomable whims of those great vivisectors, chance and natural selection? *Are we not Men?*

Human activity seems special, distinctive and elevated simply because it is we who are doing the judging. Or, to put it another way, the traditional elevation of humanity above the animals comes from the fact that we humans have nobody with whom to compare ourselves.

It was not always so; rather, our solitary state is something of a rarity. Between four and two million years ago several species of *Australopithecus* came and went. Two million years ago one or more species of *Homo* lived alongside one or more species of *Paranthropus*. Even as recently as 100,000 years ago as many as four distinct species of hominid lived in the world: 'modern' *Homo sapiens* in Africa and the Middle East; Neandertals in Europe and Central Asia; other forms of *Homo* in eastern Asia, and even *Homo erectus* in Indonesia – and these are just the ones we know about. But only *Homo sapiens* survives today.

I wonder how much of our view of evolution as linear and progressive is conditioned by our solitude; whether we might see things differently were *Homo sapiens* just one of a family of coexisting species like wapiti, wallabies or warblers; and if a cladistic appreciation of human diversity might come only after millions of years have elapsed, when we have diverged into several species.

It is hard to envisage a planet in which several different species of humanity could coexist without strife even if they shared the same language. Species, tribes, families – all are matters of degree rather than of kind, and recent history illustrates the propensity for mutual antagonism or destruction both when the protagonists are close relatives (Serbs and Croats, Hutu and Tutsi) or more distant (human beings versus the orang-utan, gorilla and chimpanzee in their shrinking forests). When white settlers reached Tasmania, they considered the stone-age aborigines as animals and hunted them down. The record is not encouraging.

Given the state of the fossil record it is possible that *Homo sapiens* still shares the Earth with other species of hominid. I do not mean the ambiguous and fleeting reports of the yeti of the Himalayas or the sasquatch of North America. Instead, I'd like to imagine the authenticated discovery of relict populations of ancient hominids alive today and what our attitude would be towards such populations.

Historians call such exercises in alternative histories 'counter-factuals'. They project how history might have been affected by an alteration in the accepted course of past events. How would things have gone with Cuba had John F. Kennedy not been assassinated? How would the course of World War II have been changed, had the Nazis pressed their advantage at Dunkirk?

Science fiction has a long tradition of alternative histories. My favourite is *The Alteration* by Kingsley Amis, a story set in 1970s Britain but in a reality in which Arthur, the first-born son of Henry Tudor and elder brother of the apostate Henry VIII, did not die young, the Reformation never happened and Martin Luther became pope. In *Virtual History*, historian Niall Ferguson explains that for this kind of exercise to be academically valid and transcend mere entertainment, documentation already available might be used to constrain events.[126] For example, even though the Nazi occupation of Russia was fleeting and incomplete, detailed plans for the future Nazi administration of Russia exist and can be studied.

My own counterfactuals – I present two below – may fall just short of perfectly valid academic exercises in alternative history. However, they illustrate that even were we to share the Earth with relict populations of ancient hominids, we would not seek to converse with them.

It is remarkable that even at the end of the twentieth century, when you would think that people had looked behind every bush and tree in every part of the world, new species of large mammal continue to be found. In 1993, zoologists described a hitherto unknown variety of antelope, the Vu Qiang 'ox' or false oryx (*Pseudoryx nghetinhensis*), from the rugged Annamite Mountains on the Lao–Vietnamese border.[127] Indochina has a habit of producing large mammals unknown to science. Another ox-like creature, the kouprey (*Bos sauveli*), was discovered there in 1937.

Perhaps other large mammals await discovery in the inaccessible Annamites. It is thought that *Homo erectus* was living in South-East Asia as recently as 100,000 years ago. Let us say that in the near future (some time around 2004) a relict population of *Homo erectus* is found in the Vu Qiang reserve created to protect the false oryx. A year later, another population is discovered by loggers in Sumatra: the *Orang Pendek* of local folklore. These creatures are found to walk as erect as you or I, use fire and make tools but have no language – at least, none we recognise – and do not wear clothes. Environmentalists and concerned governments would dither about the conservation status of something so nearly human, yet not human enough to have the power of speech. In the Annamites, the creatures would be killed by proximity to *Homo sapiens*, unintentionally wiped out by influenza. The last one, an old, solitary male, would be captured and die in a Hanoi zoo in 2018. The cause of death would be extreme old age, the animal having just passed its thirtieth birthday.

The fate of the Sumatran population would be less gentle, if just as final. The creatures would be driven out by logging, perhaps even hunted for sport. Mothers would be shot so that babies could be shipped to wealthy fanciers of exotic pets. Elsewhere in Asia there would be a thriving market for medicines prepared from the gall bladders or ground-up molars of *Homo erectus*. By 2022, the market would collapse due to lack of supply, for these ancient Sumatrans would have all gone. If *Homo erectus* lacked the power of speech, would we not think of it as an animal?

It is thought that the Neandertals survived in Spain until as recently as 33,000 years ago. Spain is a large and mountainous country. One can imagine Neandertals living around Mount Mulhacén in the Sierra Nevada until recent times. Unnoticed by the Romans, Vandals and Goths, treated as picturesque indigenes by the Caliphate, the last ones were rooted out and exterminated as devils in 1492 when Ferdinand and Isabella finally conquered the Kingdom of Granada, the last Islamic outpost in Spain. Rumour reached the Castilian court that King Muhammad XI employed 'Wild Men of Mulhacén' as his bodyguards on account of their taciturnity and immense strength. (There is some extremely suggestive evidence that Neandertals were physically capable of speech, though whether they used language is another matter.) That the Arabs employed 'demons' as bodyguards was, of course, sufficient justification for the Castilians to invade. It is not surprising that this extinction went unnoticed given that 1492 was such a busy year in which the inquisitorial monarchs expelled the Jews from their kingdom and also sponsored a successful voyage by an adventurer named Columbus.

The preceding few paragraphs are pure invention but it would be no surprise that if such relict populations were discovered, they would suffer at our hands. If they were mute, we would exterminate them as animals; if they could talk, we would subjugate them as inferior. The disappearance of the Neandertals, in particular – in real life, that is, distinct from my fantasy – has often been seen as the result of direct and possibly violent interaction between Neandertals and modern humans. But it need not have been like that.

The Neandertals had lived in Europe for hundreds of thousands of years. Modern humans arrived in Neandertal Europe around 40,000 years ago. By 30,000 years ago, the Neandertals had vanished from the record. Given the length of time that the Neandertals had been in residence, 10,000 years looks less like a long sunset than a catastrophic collapse. People have debated long and hard whether the Neandertals fell in a kind of continent-wide bloodbath. Did modern humans hunt down Neandertals as vermin in the same way that European settlers ethnically cleansed Tasmania of its indigenes in the nineteenth century? Did Neandertals perish by slow 'marginalisation' when faced by the 'superior' culture of modern humans? Or were Neandertals driven to extinction by inter-marriage, a fate likely to befall many small ethnic groups in a wider world?

These options betray our limited imaginations for they all come from recent history, analogies drawn from historical processes of colonisation and conquest that took a few decades to be played out, a century or two at most. Such analogies simply wither before the scale of Deep Time. Ten thousand years passed between the time that modern humans set foot in Europe and the Neandertals expired, an interval twice as long as recorded history. Given such a scale, you have licence to imagine possibilities far more interesting than fretful human squabbles. I should like to think that Neandertals became extinct and Moderns triumphed without their ever having met at all.

An example of this kind of slow, peaceable colonisation is the population of Britain by the North American grey squirrel (*Sciurus carolinensis*) at the expense of the native red squirrel (*Sciurus vulgaris*).[128] Since its introduction in the eighteenth century, the grey squirrel has become a familiar parkland pest, whereas the red squirrel is now endangered and confined to a few far-flung, upland areas. How did the grey squirrel make such a clean sweep? Did it interbreed with the red squirrel, swamping it with vigorous, superior grey-squirrel genes? There are no signs of this having happened. Perhaps the vigorous grey squirrel bested the shy red squirrel in open combat? The landscape is not conspicuously adorned with the products of intersciurine strife.

It seems that the grey squirrel triumphed by virtue of two quirks of physiology and behaviour. First, the grey squirrel tends to have, on average, more offspring than the red squirrel. The result is that, over time, a patch of ground inhabited by grey squirrels will become more crowded than an equivalent area patronised by red squirrels. Second, grey squirrels that move into an area tend to hold their ground, establishing themselves and breeding, spreading into adjacent areas by weight of numbers. Red squirrels, in contrast, are more shy and will move on if disturbed.

By analogy, it could have been that modern humans tended to live in denser, more interconnected populations than Neandertals, raised slightly more offspring per unit area and tended to establish more secure homesteads. If so, the Neandertals would have lost the slow, 10,000-year contest even before it started, without any arms being raised in defence or in anger. *Homo sapiens* would have emerged victorious from this long, unfought war to become the Last of the Hominids, unaware that they had been anything other than unique.

The reality, now, is that we are alone and though language is merely a thin veneer, it is also responsible for so much that we treasure, so much that makes life worth living and, at the very least, it gives us the means to talk about the problem. And yet, as I suggested above, how much *richer* our humanity might seem if we had *others* with whom to talk, compare notes, discuss our similarities and differences? For I suspect that it is the very uniqueness of human beings, our distance from the apes and the impossibility of really understanding the hominids of the past that have constrained our view of evolution that is linear and progressive. At the same time, these factors have prevented us from developing a truly comparative biology of humanity, such that we can understand what being human really means.

Human history, as we have seen, bodes ill for such interactions. The animal kingdom offers nothing with which we can hold a conversation sufficiently interesting to allow the comparison of notes on our views of existence. Some have suggested that it would take contact with message-sending extraterrestrials to shake us out of our complacency and allow us to start forming mature views of ourselves in the context of other species in the Universe. Even then, we could not know that the aliens were any more alive than a talkative automatic teller machine. Aliens, if and when we find them, could be *so* alien, *so* different from humanity, as to undermine the meaning of any exchange we might have or even to make such exchange impossible.

Even so, humanity is now at a crossroads, on the verge of moving out into space. Our steps are tentative and halting at present – despite the hoopla, the Space Station now under construction is but a small and humble thing – but we are in no hurry. We may not have the money but what we do have in abundance is time, all the time in the world, Deep Time. All we need to do is relax, hold out our hands and take it, in all its uncountable millions of years. With sufficient time, our journey into space will take us towards the comparative biology we have never had and give us the perspective we need to help us understand what it means to be human, in the sense of a species among millions, against a dark background punctuated by innumerable stars. Here, then, is my last counterfactual. Or rather, it is a vision of the far future, perhaps as remote from us now as are the Neandertals or even *Ardipithecus ramidus*.

I am writing this book in a century-old house in a western suburb of London named Ealing. From my upstairs window I can see dozens of houses, just like this one, each with its small patch of garden. A hundred years ago, none of these houses were here. Two hundred years ago, this part of the world was open farmland. A few thousand years ago, I could have sat at my window and seen an unbroken, dense forest. Twenty thousand years ago, Ealing was like Greenland, bleak and bare, a few small plants huddled against the chill blast from a continental ice sheet a few hundred miles to the north.

Fifty thousand years ago, Ealing was a prairie; there were no trees, but the grassland was rich and supported bison in herds so vast that I could have watched them from here at sunrise and still be watching them pass three sunrises later. A hundred and twenty thousand years ago, Ealing was tropical. I could have walked from here to watch hippos wallow in the Thames at Brentford, elephants graze on the banks and lions stalk oxen in the rushes. Times change and, having changed, change again. Who knows what fate Deep Time has in store for my corner of the world? In a few million years, Ealing could be a tropical beach; a range of jagged mountains; buried under a glacier a mile thick; or submerged under the sea. Given enough time, Ealing might be all of these things.

To our descendants, whoever they are and whatever they might look like, this will hardly matter. One day, be it ten thousand or ten million years hence, they will live among the planets and the stars, on planetary surfaces, in artificial space stations or in hollowed-out asteroids. Each human colony will be an inoculum of Old Earth, an island unto itself. Sufficiently dispersed to make transport difficult and genocidal strife impossible, the human population in each one will evolve in its own way. Some colonies will prosper – others will wither and die. Such is the way of things.

The time will come when our part of the Galaxy will be thinly populated with hundreds of different species of humanity, each with its own evolutionary trajectory, physical features, traditions, customs and language. Each population will have evolved in its own way but they will still share one unifying characteristic – their human heritage. Ultimately, they will all have descended from us, the residents of the Earth. *We* will be the node, for *we* will be the common ancestors of the great Clade of Humanity, a clade defined by its terrestrial origin: and thanks to the communal memory permitted by language and writing, we can be sure that it will be *our* descendants who will write the book.

Because of this common heritage, our diverse descendants will still find communion in a way that might never be achieved with extraterrestrial species. Yet these scattered populations will be sufficiently different from one another for each to have its own perspective to offer on the nature of the things that we hold dear – our place in nature, the meaning of life and intelligence, and so on. The narrow path between similarity and difference might, perhaps, be broad enough for our descendants to discover what, if anything, it means to be human.

Perhaps one of my remote descendants will make a pilgrimage to the old home planet. Its party will touch down in a parched equatorial desert that was once Ealing. Perhaps my descendant of unguessable countenance will be scaling a low, pebbly ridge when, looking down, it will chance to see a desiccated fragment weathering out of the sediment, a sliver of bone that might have been part of me.

Picking up this fragment in its hand (or claw, or tendril, or robot arm), the voyager will have cause to muse on its past, its own history in Deep Time, and wonder whether or not this fragment belonged to its distant ancestor.

And it will have no way of knowing, one way or the other.

Notes

Introduction

1 'Popular views of science assume that cause, effect and purpose can be easily discerned ...' my favourite was a news story in the spring of 1999 in which an unusual fossil human skull was claimed to have been from the offspring of interbreeding between Neandertals and anatomically 'modern' humans. How did the researchers *know* this? Did they go back in time and interview the parents of the individual whose parentage was the subject of such unseemly speculation?

2 'John McPhee, an eloquent writer on geology, coined the term "Deep Time" to distinguish geological time from the scale of time that governs our everyday lives.' The term first appears in McPhee's book *Basin and Range* (Farrar Straus & Giroux, 1981).

3 'Walter Alvarez, for example, suggests thinking of an interval of a million years as if it were a kind of geological "year".' See Alvarez's *T. rex and the Crater of Doom* (Princeton University Press, 1997).

4 'As Stephen Jay Gould has demonstrated, such misleading tales are part of popular iconography.' See the introduction to *Wonderful Life* (W. W. Norton, 1989).

5 '... would the outcome have been different from what we see, as Gould argues, or very much the same?' This – sameness – is the option preferred by

NOTES

Simon Conway Morris in *Crucible of Creation* (Oxford University Press, 1998), a work trenchantly antithetical to Gould. I have to mark my card now and say that I am more with Gould than Conway Morris, although this is more because of my unease with Conway Morris's deeply personal reasoning than for any flaw in Gould's argument.

6 'Nobody should be afraid to ask a silly question.' J. William 'Bill' Schopf of the University of California, Los Angeles (palaeontologist and author of *Cradle of Life*: Princeton University Press, 1999), has a wonderful antidote to authority. He runs a discussion group on evolution which meets every Wednesday night, under the auspices of CSEOL (Center for the Study of Evolution and the Origin of Life). The discussion is led by whoever happens to be in town that night, and can be on any aspect of science at all, from the origin of the Universe onwards. Admission is faculty only – no students are allowed. This means that people are free to ask questions without fear of being held up to ridicule by their students. The unwritten CSEOL rule is that questions can be asked at any time, and not left until the end. This spirit of free enquiry, aided by spirits of another nature, produces an atmosphere of good-natured and occasionally boisterous revelry far from the stuffy, stilted lecture-room presentation typical of most expositions of science. As a member of the audience, you feel that the ancients must have done science this way. As a lecturer at one of Bill's Wednesday nights (and I have been so privileged) you feel like a succulent Christian thrown to well-fasted lions. Only the most determined presentation gets past the first few slides.

7 'Observers innocent of science, ignorant of religious or cultural tradition and incapable of imagination would no doubt see fossils only as rocks.' Actually, this applies to any palaeontologist going into the field for the first time. As can be seen in Chapter 1, which details my visit to a palaeontological field crew in Kenya, it takes experience to train one's eyes to spot fossils. On the first day, you see nothing, not even those fossils lying right at your feet. Meave Leakey of the National Museums of Kenya has an anecdote about how we see what we are trained to see. Most people coming out to Kenya are vertebrate palaeontologists; that is, they look for the remains of backboned animals – bones and teeth. And that is what they find. Once, the crew played host to a famous expert on fossil land snails. 'I don't know how you guys can find all these bones', he complained after a few fruitless hours, 'all I can see are snails all over the place'. 'Snails?' responded the incredulous crew, who

had been on the exposure for some time without seeing a single snail shell. Seek, and ye shall find.

Chapter 1 – Nothing Beside Remains

8 'In the 1980s, Meave and her colleagues . . . surveyed dozens of potential fossil-collection sites for a hundred kilometres up and down the western shore of Lake Turkana.' The details can be found in a technical report by John Harris, Frank Brown and Meave Leakey entitled 'Stratigraphy and palaeontology of Pliocene and Pleistocene localities west of Lake Turkana, Kenya', published as *Contributions in Science* no. 399 (Natural History Museum of Los Angeles County, 1988).

9 'They include KNM-WT 15000, a near-complete skeleton of a young male *Homo erectus*, found buried under a tree-root at a place called Nariokotome.' The story of the 'Nariokotome Boy' is told by Alan Walker and Pat Shipman in their book *The Wisdom of Bones* (Weidenfeld & Nicolson, 1996).

10 '. . . in the 1980s, the Leakeys and their colleagues had found KNM-WT 17000, the so-called "Black Skull", an impressive specimen of *Paranthropus aethiopicus.*' The Black Skull was originally described as *Australopithecus boisei* by Alan Walker, Richard Leakey, John Harris and Frank Brown in '2.5-Myr *Australopithecus boisei* from west of Lake Turkana', *Nature*, vol. 322 (1986), pp. 517-22.

11 'The earliest known hominid is *Ardipithecus ramidus*, whose remains were buried in Ethiopia 4.4 million years ago.' *Ardipithecus ramidus* was originally described as *Australopithecus ramidus* (the generic name was revised later on) by Tim White, Gen Suwa and Berhane Asfaw in '*Australopithecus ramidus*, a new species of early hominid from Aramis, Ethiopia', *Nature*, vol. 371 (1994), pp. 306-12. It was accompanied by a paper by Giday WoldeGabriel and colleagues entitled 'Ecological and temporal placement of early Pliocene hominids at Aramis, Ethiopia', *Nature*, vol. 371 (1994), pp. 330-1, and a commentary by Bernard Wood 'The oldest hominid yet', *Nature*, vol. 371 (1994), pp. 280-1.

12 'A little over four million years ago, another species, *Australopithecus anamensis*, lived in the Turkana basin. Meave and her colleagues found its remains to the south and east of Lake Turkana.' *Australopithecus anamensis* was first described by Meave Leakey and her colleagues in 'New four-million-

year-old hominid species from Kanapoi and Allia Bay, Kenya', *Nature*, vol. 376 (1995), pp. 565-71.

13 '*Paranthropus boisei*; unveiled as "Nutcracker Man" or "Zinj"; *Homo habilis*, more familiarly "Handy man" – all come from the Rift.' *From Lucy to Language* by Donald Johanson and Blake Edgar (Simon & Schuster Editions, 1996) contains descriptions of nearly all the hominids mentioned in this book. A detailed account of hominid discovery, it is superbly illustrated with photographs of all the important specimens.

14 'The search for human origins in the Rift is synonymous with the name of Leakey.' *Ancestral Passions* by Virginia Morell (Touchstone, Simon & Schuster, 1996) is a comprehensive biography of the Leakey family.

15 '. . . there are few events in British history which can be treated as fact rather than conjecture.' See Peter Hunter Blair *An Introduction to Anglo-Saxon England*, (reissued 2nd edition, Cambridge University Press, 1995), p. 2.

16 'Clarke, recollecting the statement many years later.' Personal communication, 1998.

17 '. . . there is more than one way of arranging three (or more) participants in a diagram like this, so the statement about relationships in Figure 3 can be tested against possible alternatives.' Actually, there are four ways to arrange three things in such a diagram. One is the arrangement in Figure 3, in which Fred and Marmite are more closely related to each other than either is to myself. A second arrangement is shown in Figure 4, with myself and Fred more closely related to each other than either is to Marmite. A third arrangement is similar – just swap Marmite and Fred around. The fourth arrangement is one in which we all branch from a single node. This is called an 'unresolved trichotomy' and reflects a situation in which none of the three other arrangements can be chosen unequivocally over any other.

Chapter 2 – Hunting Unicorns

18 Larry Gonick's *Cartoon History of the Universe* is published by Rip-Off Press, San Francisco.

19 '[Lungfishes] live in fresh water, but when the water dries out, the African and South American forms can cocoon themselves in burrows until the rains

come again, in the manner of some desert amphibians.' Per Erik Ahlberg reminds me that the Australian lungfish does not have this habit, but lives in permanent water bodies.

20 'The most recent coelacanth known exclusively from fossils died out around 70 million years ago, when dinosaurs still ruled the Earth.' See Peter Forey, *The History of the Coelacanth Fishes* (Chapman & Hall, 1998) for a comprehensive, scholarly account of the group.

21 'There seemed no reason to think otherwise until the carcass of a recently dead coelacanth was recovered off South Africa in 1938, after a gap of 70 million years.' See K. S. Thomson's *Living Fossil: The Story of the Coelacanth* (W. W. Norton, 1992) for a lively account of the discovery of the living coelacanth.

22 'The discovery of another population 10,000 kilometres away, off Sulawesi in Indonesia in 1998, was the zoological surprise of the year.' See Mark V. Erdmann *et al.*, 'Indonesian King of the Sea discovered', *Nature*, vol. 395 (1998), p. 335, with the accompanying commentary by Peter L. Forey, 'A home from home for coelacanths' *Nature*, vol. 395 (1998), pp. 319-20.

23 'Jarvik's monograph on *Ichthyostega* finally appeared in 1996, a short time before Jarvik's own death.' Jarvik's technical monograph 'The Devonian Tetrapod *Ichthyostega*', *Fossils and Strata*, vol. 40 (Scandinavian University Press, 1996), pp. 1-212 contains an illuminating first-hand account of Arctic fieldwork in the 1930s.

24 'A critical look at *Ichthyostega* is something like the study of the unicorn by Han Yu, a ninth-century Chinese writer, as dubiously reported by Jorge Luis Borges in his essay *Kafka and his Precursors*.' (See *Kafka and his Precursors*, translated by J. E. Irby, in the collection *Labyrinths*: first published by New Directions Publishing Corporation, 1962: I cite the Penguin edition of 1970.) 'Dubious', because although some of Borges's literary criticism is genuine, some of his essays are fictions cast in the form of criticism to make a point about the derivative nature of authorship: this is precisely the point made in *Kafka and his Precursors*.

25 'Jenny and Mike published a series of papers on *Acanthostega* through the late 1980s and 1990s, painting a progressively more complete portrait of *Acanthostega*, a tetrapod even stranger than *Ichthyostega*.' Carl Zimmer has written about the subject in his book *At the Water's Edge* (Free Press, 1998).

Mike Coates gives a comprehensive account of work on *Acanthostega* in 'The Devonian tetrapod *Acanthostega gunnari*, Jarvik: postcranial anatomy, basal tetrapod interrelationships and patterns of skeletal evolution', *Transactions of the Royal Society of Edinburgh: Earth Sciences*, vol. 87 (1996), pp. 363-421. For a review on early tetrapod evolution, see Per E. Ahlberg and Andrew R. Milner, 'The origin and early diversification of tetrapods', *Nature*, vol. 368 (1994), pp. 507-14.

26 '. . . all these details of pteraspid anatomy, and more, demonstrate a close link with lampreys – even though lampreys lack the distinctive head armour of pteraspids.' For up-to-date views on the relationships of lampreys, hagfishes and extinct jawless fishes such as pteraspids, see Peter Forey and Philippe Janvier, 'Evolution of the early vertebrates', *American Scientist*, vol. 82 (1994), pp. 554-65; Peter Forey, 'Agnathans recent and fossil, and the origin of jawed vertebrates', *Reviews in Fish Biology and Fisheries*, vol. 5 (1995), pp. 267-303; Philippe Janvier, 'The dawn of the vertebrates: characters versus common ascent in the rise of current vertebrate phylogenies', *Paleontology*, vol. 39 (1996), pp. 259-87; and Philippe Janvier's splendid volume *Early Vertebrates* (Oxford: Oxford University Press, 1996).

27 'We do not know how many digits *Elginerpeton* wore on its limbs. It could have been three, nine, six, five, or some other number. Whatever it turns out to be, there is no longer any reason to think that it *must* have been five.' See Per Ahlberg, 'Postcranial stem tetrapod remains from the Devonian of Scat Craig, Morayshire, Scotland', *Zoological Journal of the Linnean Society*, vol. 122, pp. 99-141.

28 'As long ago as 1938, a British researcher, Stanley Westoll, wrote a short report in *Nature* on a skull roof of a Devonian lobe-finned fish called *Elpistostege* which he claimed was a tetrapod.' See T. Stanley Westoll, 'Ancestry of the Tetrapods' *Nature*, vol. 141 (1938), pp. 127-41. This paper was all about the arrangements of bones in the skull roofs of fishes and early tetrapods, a field in which Westoll was an acknowledged expert. Looking at this paper reminds me of when I went to my first palaeontological conference, hustling for doctorate opportunities and meeting luminaries in the field. It was 1983, the same year that I worked for the summer at the Natural History Museum. Westoll, although old and frail, was very much in attendance and had a commanding presence. I had spent much of the summer poring over

one of Jarvik's monographs, and had been confused by his use of skull-bone nomenclature. One evening at dinner, in all innocence, I asked Westoll to enlighten me – a hush descended on the entire refectory table as Westoll gave an impromptu tutorial to me and the other diners.

29 'A recent cladogram (Figure 8) of lobe-finned fishes and tetrapods, drawn up by Per Ahlberg and Zerina Johanson of the Australian Museum in Sydney, says more about the emergence of tetrapods than any amount of storytelling.' See Per Ahlberg and Zerina Johanson, 'Osteolepiforms and the ancestry of tetrapods', *Nature*, vol. 395 (1998), pp. 792-4.

30 'Another specialised group of lobe-finned fishes was the rhizodonts. These were large animals ... with digit-like structures within their fins.' See E. B. Daeschler and N. Shubin, 'A fish with fingers?' *Nature*, vol. 391 (1998), p. 133.

31 '... two men, Othniel Charles Marsh of Yale, and Edward Drinker Cope of Harvard, whose energy, rivalry and mutual hatred led to ever greater frenzies of collection, as each strove to outdo the other.' *The Life of a Fossil Hunter* by Charles Sternberg, a fossil hunter who worked for Cope, is a thrilling first-hand account of fieldwork in the Wild West. The book was originally published by Holt in 1909, but was reissued in 1990 by Indiana University Press.

32 'Because of this uncertainty, you cannot assume without question that they were used for the same purposes as teeth are used for today, such as biting and grasping.' The first report of the discovery of the conodont animal with cautious speculations on its affinities was 'The conodont animal' by D. E. G. Briggs, E. N. K. Clarkson and R. J. Aldridge, in *Lethaia*, vol. 16 (1983), pp. 1-14. For other recent works on the lives and times of conodonts, see Mark Purnell, 'Microwear on conodont elements and macrophagy in the first vertebrates', *Nature*, vol. 374 (1995), pp. 798-800; S. E. Gabbott, R. J. Aldridge and J. N. Theron, 'A giant conodont with preserved muscle tissue from the Upper Ordovician of South Africa', *Nature*, vol. 374 (1995), pp. 800-3; Philippe Janvier, 'Conodonts join the club', *Nature*, vol. 374 (1995), pp. 761-2; R. J. Aldridge *et al.*, 'The apparatus, architecture and function of *Promissum pulchrum* Kovács-Endrödy (Conodonta, Upper Ordovician) and the prioniodontid plan', *Philosophical Transactions of the Royal Society of London*, Series B, vol. 347 (1995), pp. 275-91; and R. J. Aldridge *et al.*, 'The anatomy of conodonts', *Philosophical Transactions of the Royal Society of London*, Series B, vol. 340 (1993), pp. 405-21.

33 'We could be the starfishes that lost their calcite skeletons.' My book *Before the Backbone* (Chapman & Hall, 1996) gives a full account of the latest theories about the origins of vertebrates, including a discussion about Jefferies' ideas.

34 'This is not a description of a real animal that once existed, but an alien from a science-fiction novel . . .' See Greg Bear, *Eternity* (Warner Books, 1994). Quote used with the kind permission of the author.

35 'Simon Conway Morris of the University of Cambridge . . . named it *Hallucigenia*, on account of its bizarre, "dreamlike" appearance.' For Simon Conway Morris's paper on *Hallucigenia* see 'A new metazoan from the Cambrian Burgess Shale of British Columbia', *Palaeontology*, vol. 20 (1977), pp. 623-40.

36 'And yet the only living representative of this group is the humble velvet worm, which lives exclusively on land. This remarkable group of animals has completely vanished from its oceanic home.' Popular accounts of *Hallucigenia* and other animals from the Burgess Shales of British Columbia can be found in Simon Conway Morris's *Crucible of Creation* (Oxford: 1998), which offers an antithetical approach to Stephen Jay Gould's *Wonderful Life* (Norton, 1989). More technical accounts can be found in Derek E. G. Briggs, Douglas H. Erwin and Frederick J. Collier, *Fossils of the Burgess Shale* (Smithsonian Institution Press, 1994).

37 '. . . a startling suggestion – that *Megatherium* was the largest mammalian meat-eater of all time.' See the paper by Richard A. Fariña and R. E. Blanco, '*Megatherium*, the stabber', *Proceedings of the Royal Society of London*, Series B, vol. 263 (1996), pp. 1725-9.

38 'In 1996, David S. McKay . . . published a report . . . that signs of past life had been found inside a meteorite discovered in Antarctica, but which had originated on the surface of the planet Mars.' The paper, from McKay and colleagues, is 'Search for past life on Mars: possible relic biogenic activity in Martian meteorite ALH84001', *Science*, vol. 273 (1996), pp. 924-9.

39 'J. William Schopf of the University of California, Los Angeles, is an authority on the fossils of bacteria and other microscopic organisms.' See Schopf's recent book *Cradle of Life* (Princeton University Press, 1999).

Chapter 3 – There Are More Things

40 'Even relativity, though, is not the last word, and it is possible that a future physicist will find a better description of the world.' For example, there is as yet no way within the theory of relativity that gravity can be expressed in terms of quantum mechanics. A successful theory of 'quantum gravity' would stand in relation to relativity as relativity does to Newtonian mechanics.

41 'L. Anders Nilsson of the University of Uppsala in Sweden, reported a test of a hypothesis proposed by Darwin in 1862 . . .' For Nilsson's paper, 'The evolution of flowers with deep corolla tubes', see *Nature*, vol. 334 (1992), pp. 147-9. See also a recent commentary by Nilsson, 'Deep flowers for long tongues', *Trends in Ecology and Evolution*, vol. 13 (1998), pp. 259-60 on some other hypotheses to explain why hawkmoths have long tongues.

42 '. . . the network of interactions in an ecosystem is so complicated that it is difficult to predict precisely what would happen were one particular participant in an ecosystem to become more or less abundant, or to change its habits.' Scientists are increasingly interested in the unpredictable behaviour of whole ecosystems. A paper by David Tilman and colleagues, 'Habitat destruction and the extinction debt', *Nature*, vol. 371 (1994), pp. 65-6, presents a startling example of this, in which the gradual loss of a habitat has the gravest effects on the most common species in the habitat, not the rarer ones, as you might have expected.

43 'A plausible yet untestable view of human origins of long standing is the "aquatic-ape" scenario.' The leading advocate of the aquatic-ape scenario is Elaine Morgan, who has discussed it in several books, including *The Scars of Evolution: What Our Bodies Tell Us About Human Origins* (reprinted in paperback by Oxford University Press, 1994).

44 'It is tempting to use present-day adaptations to explain past history, but one need have nothing to do with the other. Stephen Jay Gould and Richard C. Lewontin of Harvard University explored this tension in an influential paper.' For this paper see 'The spandrels of San Marco and the Panglossian paradigm: a critique of the adaptationist programme', *Proceedings of the Royal Society of London*, Series B, vol. 205 (1979), pp. 581-98.

45 '. . . Robert T. Bakker looks at how our images of dinosaurs have changed with scientific fashion.' For Robert T. Bakker's 'The Return of the Dancing

Dinosaurs', see *Dinosaurs Past and Present*, Vol.1 (Natural History Museum of Los Angeles County, 1987), pp. 38-69.

46 'Towns and cities would be recognizable by moving knots of roundworms in human shape, each one the catalogue of infestation in each one of us.' In his charming textbook *Animals Without Backbones*, Ralph Buchsbaum attributes this nematological nightmare to a – sadly unnamed – specialist on roundworms. So much was true, at any rate, of my well-thumbed copy from twenty years ago. I have not examined the current edition which was published in 1987 by the University of Chicago Press.

47 ' In this way, mate choice is specifically and directly influenced by genetic variation maintained as a hedge against disease.' The influence of immune recognition on mating patterns in mice was demonstrated in a paper by Wayne K. Potts and colleagues entitled 'Mating patterns in seminatural populations of mice influenced by MHC genotype', *Nature*, vol. 352 (1991), pp. 619-21; J. C. Howard wrote an intriguing commentary on this paper, 'Disease and evolution', *Nature*, vol. 352 (1991), pp. 565-7. Potts's group made the link between communal nesting patterns and immune-system genes in a paper the following year: see C. J. Manning and colleagues, 'Communal nesting patterns in mice implicate MHC genes in kin recognition', *Nature*, vol. 360 (1992), pp. 581-3.

48 'The earliest fossils are of bacteria, found in the 3.6-billion-year-old Apex Chert of Australia, first described by J. William Schopf.' See Schopf's paper 'Microfossils of the Early Archean Apex chert: new evidence of the antiquity of life', *Science*, vol. 260 (1993), pp. 640-6. Traces of life, detectable as geochemical alterations in rocks, are detectable in rocks as old as 3.8 billion years. See S. J. Mojsis *et al.*, 'Evidence for life on Earth before 3,800 million years ago', *Nature*, vol. 384 (1996), pp. 55-9, and the accompanying commentary by J. M. Hayes, 'The earliest memories of life on Earth', *Nature*, vol. 384 (1996), pp. 21-2.

49 '. . . some of the organisms preserved as fossils more than 3 billion years old are virtually identical with organisms, called cyanobacteria, living today.' See J. W. Schopf, 'Disparate rates, differing fates: tempo and mode of evolution from the Precambrian to the Phanerozoic', *Proceedings of the National Academy of Sciences of the USA*, vol. 91 (1994), pp. 6735-42.

Chapter 4 – Darwin and His Precursors

50 'Mike Benton has compiled a bibliography of hypotheses to explain dinosaur extinction, culled from the scientific literature.' See Benton's paper, 'Scientific methodologies in collision: the history of the study of the extinction of the dinosaurs', *Evolutionary Biology*, vol. 24 (1990), pp. 371-400.

51 'Much evidence supports the hypothesis that an asteroid landed at the required time and caused environmental upset of the required magnitude to have done for the dinosaurs.' See Walter Alvarez, *T. rex and the Crater of Doom* (Princeton University Press, 1997), a book that is far better than its cheesy title suggests.

52 'This view of evolution is seen everywhere in advertisements.' In *Wonderful Life* (Norton, 1981) Stephen Jay Gould shows us his impressive collection of advertisements, in which advertisers (who do no more than reflect the public consciousness) clearly see evolution as progressive.

53 The conclusion – the catch-line meant to stay in the mind of the viewer – was "It's Evolved".' It is perhaps ironic that the voice-over for this commercial was provided by a well-known geneticist.

54 'The impact of Darwin's views on modern thought has been so profound that it is extremely hard for us, today, to imagine how people thought about the history of life before the publication of the *Origin of Species* in 1859.' See David L. Hull's book, *Science as a Process* (University of Chicago Press, 1988). This book is highly recommended as an excellent overview of the development of thought in evolutionary biology, including cladistics.

55 '. . . until recently, Darwin's home, Down House in Kent, languished in slow decay for lack of funds.' Darwin is, in fact, commemorated by a rather ordinary suburban street, Darwin Road, in Brentford, west London, within sight of the elevated section of the M4 motorway. There is also a Darwin Street in Walworth, a down-at-heel district of south-east London. There is, however, a Rue Darwin in the picturesque Montmartre district of Paris. This modest street can be found close to the altogether grander Rue Lamarck. Never let it be said that city planners don't have a sense of humour.

56 'Should you find yourself at Arlanda International Airport in Stockholm, you might consider stopping for refreshment at the Bar Linné, named after

the eighteenth-century Swedish naturalist Carl von Linné, better known as Linnaeus.' Linnaeus, unlike Darwin, is no stranger to commemoration. His portrait features on Swedish banknotes, keeping company with the singer Jenny Lind, the 'Swedish Nightingale'. As far as I know, there are no plans to open a Darwin Bar at Heathrow Airport in London.

57 'The answer to Darwin's problem was the discovery, long after his death, of particles of inheritance called genes.' Darwin died in 1882.

58 'Modern, textbook views of evolutionary biology are referred to as "neo-Darwinian" and are substantially based on the views of Dobzhansky and his contemporaries.' In *Science as a Process* (University of Chicago Press, 1988), David L. Hull gives a full account of the work of Dobzhansky and his colleagues and the development of the Modern Synthesis.

59 'Mayr sought to solve this problem by setting out the "biological species concept" . . .' See Mayr's book *Principles of Systematic Zoology* (New York: McGraw-Hill, 1969).

60 'Simpson chose history and lineage and, in 1961, put forward his "evolutionary species concept" . . .' See Simpson's book *Principles of Animal Taxonomy* (New York: Columbia University Press, 1961).

61 '. . . the distinguished palaeontologist Alfred Sherwood Romer, published an evolutionary scenario that serves us here as a case history of this kind of thinking in action.' See A. S. Romer, 'Eurypterid influence on vertebrate history', *Science*, vol. 78 (1933), pp. 114-17.

62 '. . . Gibbon's *Decline and Fall of the Roman Empire*, a title that historian Niall Ferguson sees as an exemplar of a style of history told according to the conventions of narrative form, with a beginning, a middle and an end, key characters, key events, and a linear plot.' See Niall Ferguson (ed.), *Virtual History: Alternatives and Counterfactuals* (London: Picador, 1997).

63 'Virtually every group of non-vertebrate animal has been proposed as the direct ancestor of the vertebrates.' See my book *Before the Backbone* (Chapman & Hall, 1996) for a historical survey of ideas about vertebrate origins, and assessments of current ideas.

64 'By extension, fishes had to have evolved from those creatures living in the preceding Silurian Period which had the most complex nervous systems,

creatures such as the horseshoe crab.' The discussion between Gaskell and other participants in the debate at the Linnean Society was transcribed and published as 'Discussion on the origin of vertebrates', *Proceedings of the Linnean Society of London*, Session 122 (1910), pp. 9-50.

Chapter 5 – The Gang of Four

65 '[Gardiner] would sooner chat over a pint about . . . his almost obsessional interest in the famous Piltdown Man hoax of 1912.' In 1912, a Sussex-based antiquary named Dawson announced the discovery of a hominid skull, jawbone and associated stone tools in a gravel pit at Piltdown, near Newick in Sussex, England. The discovery was a hoax, but its fraudulent nature was exposed only in the 1950s, after patient forensic work by Kenneth P. Oakley and colleagues, using new methods of dating artefacts. The bones and tools were of recent date, stained and abraded to look old. However, the identity of the hoaxer has never been securely established and the fraud has been blamed on many scientists of the day, as well as the French Jesuit and palaeontologist Teilhard de Chardin, the writer and spiritualist Arthur Conan Doyle, and Dawson himself, working either alone or with an accomplice. In the 1950s, Gardiner was a young researcher who knew Oakley and his colleagues, the people working to expose the fraud. Over time, he became convinced that the hoaxer was Martin Hinton, the authority on fossil rodents at the Natural History Museum. Hinton was a skilled technician; knew the Piltdown site; had a grudge against senior Museum scientist Arthur Smith Woodward, who was taken in by the Piltdown hoax; and had left memoirs virtually confessing his involvement in the affair. Gardiner was not alone in his suspicions of Hinton. Quite independently, Andrew Currant – a curator at the Museum and, like Hinton, an expert on fossil rodents – came across an old travelling trunk, found in a Museum attic during routine roof repairs. The trunk was marked with Hinton's initials. Inside the case, beneath dozens of glass vials of rodent remains, were assorted bones, stained and filed in much the same way as the Piltdown specimens – as if they were practice runs for the forgery. Currant brought his finds to Gardiner's attention, and Gardiner then told me (I wrote up the story in *Nature*, vol. 381 (1996), pp. 261-2).

66 'However, it took me a little while before I realised . . . why other scientists stayed away from the Cranley as if there was a curse on it.' The Cranley

closed down in the mid-1980s to be reopened as a yuppie wine bar. 'There'll be VW Golfs parked seven deep outside', Forey remarked to me somewhat ruefully. The cladists moved their lunch spot to a pub called the Rose. Although the Rose was more salubrious and comfortable than the Cranley, the cladists had lost their spiritual home.

67 'In the article, the authors argued that lungfishes (those curious, semi-amphibious fishes found in South America, Africa and Australia) were the closest living relatives of the tetrapods.' For this article by Rosen and colleagues, see 'Lungfishes, Tetrapods, Paleontology and Plesiomorphy', *Bulletin of the American Museum of Natural History*, vol. 167 (1981), pp. 163-275.

68 'The possession of choanae suggested that tetrapods and lungfishes share a common ancestry that excludes other fishes, in the same way that the shared possession of pointed ears and whiskers is evidence that my cats Marmite and Fred have a common ancestry that excludes me.' Later work has shown that the Gang of Four's thesis was flawed as regards some of its details. Subsequent cladistic analyses, for example that of Ahlberg and Johanson (see Figure 8), show that tetrapods are now believed to be more closely related to *Eusthenopteron* than to lungfishes. Such is progress – but it does not diminish the philosophical point that the Gang of Four was making about the validity, or otherwise, of evolutionary scenarios based on inferences about ancestry and descent. The most we can know of *Eusthenopteron* is that it is a cousin, its status discovered by the accumulation of evidence, its phylogenetic position supported by the testing of cladograms, each of which is a hypothesis and therefore provisional; *Eusthenopteron* is emphatically not an ancestor, its status determined by adaptive reasoning after the fact, its phylogenetic position – as an ancestor – justified only by assertion.

69 'It was only later on that phylogenetic systematics came to be called "cladistics".' It was Ernst Mayr, author of the biological species concept, who coined the term 'cladistics' (for details on the early interaction between the cladists and the traditional school of evolutionary biology, see David Hull's book *Science as a Process*, University of Chicago Press, 1988).

70 'As Nelson put it, rhetorically, in a lecture in the early 1970s – "Do the rocks speak?"' The source for this quotation is an unpublished manuscript for a lecture by Nelson, believed to have been delivered around 1971. I am grateful to Peter Forey for drawing it to my attention.

71 'Nelson shared his new enthusiasm with Colin Patterson, who was also mulling over the implications of Hennig's ideas.' Nelson's own account of the development of cladistics and Colin Patterson's part in it can be found in a celebration of Colin Patterson's life, published in *The Linnean*, the magazine of the Linnean Society of London (1999). Nelson's account is rather different from that given by David Hull in his book *Science as a Process* (University of Chicago Press, 1988). Hull suggests that Nelson was the first to tell Patterson about cladistics. Nelson shows, in contrast, that Patterson had already been thinking over Hennig's ideas.

72 'Whenever the correspondence looked like flagging, *Nature*'s leader writers kept the pot boiling with a few playful thunderbolts of their own ... if Bev wanted publicity, he got it.' The full Halstead-inspired correspondence can be found in *Nature*, vol. 276 (1978), pp. 759-60; vol. 277 (1979), pp. 175-6; vol. 280 (1979), pp. 541-2; vol. 288 (1980), pp. 208, 430 and 638; vol. 289 (1981), pp. 8, 105-7, 735 and 142; vol. 290 (1981), pp. 75-6, 82, 286 and 730; vol. 291 (1981), pp. 7-8, 104, and vol. 292 (1981), pp. 1-2, 95-6, 395-6 and 403-4.

73 'Halstead wrote ... *Homo sapiens*.' See *Nature*, vol. 288 (1980), p. 208.

74 'Colin Patterson replied directly to this passage ... judge the evidence.' See *Nature*, vol. 288 (1980), p. 430.

75 'Donn Rosen added his own voice to the debate.' See *Nature*, vol. 289 (1981), pp. 8 and 105.

76 '... cladistics has become a universal tool for understanding matters as diverse as ... the evolution of languages and literature'. See N. I. Platnick and H. Don Cameron, 'Cladistic methods in textual, linguistic and phylogenetic analysis', *Systematic Zoology* (1977), pp. 380-5.

77 'They read the history of roundworms from modern species, or, to be precise, their molecules.' See Mark L. Blaxter and colleagues, 'A molecular evolutionary framework for the phylum Nematoda', *Nature*, vol. 392 (1998), pp. 71-5.

78 '... the complete gene sequence, or "genome", of [*C. elegans*] ... is now known'. See 'Genome sequence of the nematode *Caenorhabditis elegans*. A platform for investigating biology', *Science*, vol. 282 (1999), pp. 2012-18.

79 '. . . the Léopoldville virus was probably close to the original source of the epidemic in time and space.' See Tuofu Zhu and colleagues, 'An African HIV-1 sequence from 1959 and implications for the origin of the epidemic', *Nature*, vol. 391 (1998), pp. 594-7.

80 'More recent molecular sequence comparison, by Beatrice H. Hahn of the University of Alabama at Birmingham, Alabama and colleagues, has followed HIV-1 back to its original source, among primates . . . whose geographic range coincides with the most divergent – and thus most ancient – lineages of HIV-1 in humans.' See Feng Gao and colleagues, 'Origin of HIV-1 in the chimpanzee *Pan troglodytes troglodytes*', *Nature*, vol. 397 (1999), pp. 436-41.

81 'Using this information, they created a new cladogram for swordtails that showed how the sword is repeatedly gained and lost in evolution: and that the common ancestor most likely had a sword.' See Axel Meyer, Jean M. Morrissey and Manfred Schartl, 'Recurrent origin of a sexually selected trait in *Xiphophorus* fishes inferred from a molecular phylogeny', in *Nature*, vol. 368 (1994), pp. 539-42, and the accompanying commentary by Andrew Pomiankowski, 'Swordplay and sensory bias', *Nature*, vol. 368 (1994), pp. 494-5.

82 'The bacteria used instructions carried in the artificial genes to make pancreatic ribonuclease . . . in quantities that could be harvested and tested.' For details of this work, see Thomas M. Jermann, Jochen G. Opitz, Joseph Stackhouse and Steven A. Benner, 'Reconstructing the evolutionary history of the artiodactyl ribonuclease superfamily', *Nature*, vol. 374 (1995), pp. 57-9, and also the accompanying commentary by Caro-Beth Stewart, 'Active ancestral molecules', *Nature*, vol. 374 (1995), pp. 12-23.

83 'Some recent studies suggest that hippos are more closely related to whales than they are to other even-toed ungulates such as ruminants or other pigs.' For a quick guide to this controversial area, see the short commentary by M. C. Milinkovitch and J. G. M. Thewissen, 'Even-toed fingerprints on whale ancestry', *Nature*, vol. 388 (1997), 622-3.

84 'Cladistics, because it assumes much less about the evidence, reveals a great deal more': a line I stole with subtle daring (and subsequently paraphrased) from Mark Norell of the American Museum of Natural History in New York.

85 'Adrian C. Barbrook from the University of Cambridge and his colleagues applied cladistics to understanding the relationships of 58 extant fifteenth-century manuscripts of the Prologue to the *Wife of Bath's Tale* . . .' See Barbrook *et al.*, 'The phylogeny of *The Canterbury Tales*', *Nature,* vol. 394 (1998), p. 839.

86 '. . . linguists are increasingly teaming up with geneticists to trace how the evolution of languages tracks the movement of human groups in prehistory.' See, for example, Luca Cavalli-Sforza, 'Genes, peoples and languages' *Proceedings of the National Academy of Sciences of the USA*, vol. 94 (1997), pp. 7719-24.

Chapter 6 – The Being and Becoming of Birds

87 'I shall discuss this approach in as convincing a way as I can. Later in this chapter, I shall show how and why it is wrong.' The traditional scenario in which the origin of birds is linked with the origin of flight is set out most clearly and fully in *The Origin and Evolution of Birds*, by Alan Feduccia (New Haven: Yale University Press, 1996). Feduccia has been the most active proponent of the traditional, scenario-based view of bird origins, and the most vocal opponent of the cladistic interpretation of bird evolution, in which birds are seen as close relatives of theropod dinosaurs. In this capacity, Feduccia stands in relation to the latter-day cladists rather as Halstead did to the Gang of Four.

88 '. . . there are reports of cats requiring only minor veterinary treatment after falls of more than thirty stories.' Veterinary surgeons in Manhattan even have a name – 'High-Rise Syndrome' – for the suite of injuries sustained by parachuting cats. In a report entitled 'High-rise syndrome in cats', veterinarians W. O. Whitney and C. J. Mehlhaff investigated the severity of High-Rise Syndrome as a function of height fallen by Manhattan cats, measured in number of stories. See *Journal of the American Veterinary Medical Association*, vol. 191 (1987), pp. 1399-403.

89 'The first fossil bird was discovered in 1861, two years after Darwin's *Origin of Species* was published.' The history of *Archaeopteryx* is told by Pat Shipman in *Taking Wing: Archaeopteryx and the Origin of Bird Flight* (New York: Simon & Schuster, 1998).

90 'Some of the bones have holes that could have admitted extensions for the air-sac system.' See Brooks Britt and colleagues in *Nature*, vol. 395 (1998), pp. 374-6.

91 This is as much as can be justified given the evidence, and is independent of chronology.' The objection has, in any case, been overtaken by events, as the fossil record of dromaeosaurs is no longer restricted to the later parts of the Cretaceous: there is evidence that they lived as long ago as the late Jurassic, coeval with *Archaeopteryx* (see J. A. Jensen and K. Padian, 'Small pterosaurs and dinosaurs from the Uncompahgre fauna (brushy Basin member, Morrison Formation: ?Tithonian) Late Jurassic, western Colorado', *Journal of Paleontology*, vol. 63 (1989), pp. 364-74).

92 'The question is whether bird fingers and dromaeosaur fingers are, in evolutionary terms, the "same" three fingers. If they are not, the close link between birds and dinosaurs would be weakened.' See C. A. Burke and Alan Feduccia, 'Developmental patterns and the identification of homologies in the avian hand', *Science*, vol. 278 (1997), pp. 666-8.

93 'In support of this view, they . . . claim to identify similar adaptations in fossils such as *Archaeopteryx*.' See Alan Feduccia's book *The Origin and Evolution of Birds* (New Haven: Yale University Press, 1996) for a full discussion of this point.

94 'The very existence of dromaeosaurs exposes the weakness of the foundations on which rests the entire edifice of the traditional scenario of bird evolution.' However, by the argument above, you should be wary of claims for or against particular lifestyles in extinct forms. Although dromaeosaurs look like bipedal runners, they might equally have been good swimmers, or even have climbed trees. Flight, however, is quite another matter. Golden retrievers are good swimmers without being dolphins and goats are good tree-climbers without being monkeys but neither goats nor golden retrievers can fly.

95 'Quarries in the province of Liaoning, to the north of Beijing, had yielded fossils of a primitive bird, *Confuciusornis*.' See L. Hou, L. D. Martin and Alan Feduccia, 'A beaked bird from the Jurassic of China', *Nature*, vol. 377 (1995), pp. 616-18.

96 '*Confuciusornis* joined a steadily accumulating catalogue of fossil birds, unearthed in the 1980s and 1990s from a small number of fossil sites in

China, Spain and other countries.' For recent reviews on progress in bird evolution see two articles by Kevin Padian and Luis Chiappe, 'The origin of birds and their flight', *Scientific American* (February 1998), pp. 38-47, and 'The origin and early evolution of birds', *Biological Review*, vol. 73 (1998), pp. 1-42.

97 '. . . the age of the Liaoning fossil beds has been contentious, but they are now thought to be from the early Cretaceous, about 124 million years old.' See Carl C. Swisher III, Yuan-Qin Wang, Xiao-Lin Wang, Xing Xu and Yuan Wang 'Cretaceous age of the feathered dinosaurs of Liaoning, China', *Nature*, vol. 400 (1999), pp. 58-61.

98 'The belief was enshrined in the name [Chen] and his colleagues gave the fossil: *Sinosauropteryx*, the Chinese winged lizard.' The name is slightly fanciful in that *Sinosauropteryx* had no wings and could not have flown. See Chen Pei-Ji, Z.-M. Dong and S.-N. Zhen, 'An exceptionally well-preserved theropod dinosaur from the Yixian Formation of China', *Nature*, vol. 391 (1998), pp. 147-52 and the accompanying commentary by D. M. Unwin, 'Feathers, filaments and theropod dinosaurs', *Nature*, vol. 391 (1998), pp. 119-20.

99 '. . . two more dinosaurs clothed in *Sinosauropteryx*-like fibres. One is a dromaeosaur but the other is a therizinosaur, a member of an extremely aberrant theropod offshoot.' See Xu Xing *et al.*, 'A therizinosauroid dinosaur with integumentary structures from China', *Nature*, vol. 399 (1999), pp.350-54, for a description of the 'feathered' therizinosaur, and Xing Xu, Xiao-Lin Wang and Xiao-Chun Wu, 'A dromaeosaurid dinosaur with a filamentoins integument from the Yixian Formetron of China', *Nature*, vol. 401 (1999), pp. 262-66, for a description of a very small, very hairy dromaeosaur with huge claws. At first I thought it ought to have been a kitten rather than a dinosaur.

100 'The fruits of this collaboration led to a paper on the dinosaurs *Protarchaeopteryx* and *Caudipteryx*, which was published in *Nature* in June 1998.' See Ji Qiang, P. J. Currie, M. A. Norell and Ji Shu-An, 'Two feathered dinosaurs from northeastern China', *Nature*, vol. 393 (1998), pp. 753-61.

101 'The traditional view of bird origins, as well as being theoretically indefensible, has been trounced by the fossil evidence.' Challenges to the bird–dinosaur view still come from a few quarters. However, it is significant

that none has ever been framed in terms of cladistics, being based instead on appeals to plausibility based on the interpretation of selected features of the anatomy.

102 'The arm-bones of *Unenlagia* . . . The dinosaur *Velociraptor* had a wishbone . . . There is a fossil of the dinosaur *Oviraptor*, petrified while seated on its nest . . . in a pose much like that of a farmyard chicken.' F. E. Novas and P. F. Puerta described the remarkable dinosaur *Unenlagia* in 'New evidence concerning avian origins from the Late Cretaceous of Patagonia', *Nature*, vol. 387 (1997), pp. 390-2. See also the accompanying commentary by L. M. Witmer, 'A new missing link', *Nature*, vol. 387 (1997), pp. 349-50. Mark Norell and colleagues discuss the wishbone of *Velociraptor* in *Nature*, vol. 389 (1997), p. 447. The nesting *Oviraptor* is described by Norell and colleagues in a paper entitled 'A nesting dinosaur', *Nature*, vol. 378 (1995), pp. 774-6. These are just a few examples of recent findings of bird-like features in dinosaurs.

103 'Many dinosaur skeletons . . . and many other creatures of all kinds.' See Michael Novacek's *Dinosaurs of the Flaming Cliffs* (Anchor, 1996) for a compelling summary of the history of palaeontology in the Gobi Desert and an account of recent discoveries made by the ongoing joint expedition of the American Museum of Natural History and the Mongolian Academy of Sciences.

104 'There was enough to see what kind of animal these bones belonged to and offer a formal, taxonomic description with a name.' See Perle Altangerel, Mark A. Norell, Luis M. Chiappe and James M. Clark, 'Flightless bird from the Cretaceous of Mongolia', *Nature*, vol. 362 (1993), pp. 623-6. The animal was originally called *Mononychus*, but the researchers discovered, after the original publication, that this name had already been used for a genus of beetle. Taxonomy forbids such confusing duplication, and so Altangerel and colleagues published the name *Mononykus* in a subsequent note.

105 'The latest addition to the family is a fossil called *Shuuvuia*, which, like *Mononykus*, comes from Mongolia'. See L. M. Chiappe, M. A. Norell and J. M. Clark, 'The skull of a relative of the stem-group bird *Mononykus*', *Nature*, vol. 392 (1998), pp. 275-8.

Chapter 7 – Are We Not Men?

106 'The fossils from Aramis form the basis of a new species, *Australopithecus ramidus* – later renamed *Ardipithecus ramidus* – the earliest known hominid.' See Tim White, Gen Suwa and Berhane Asfaw, '*Australopithecus ramidus*, a new species of early hominid from Aramis, Ethiopia', *Nature*, vol. 371 (1994), pp. 306-12, accompanied by a paper by Giday WoldeGabriel *et al.*, 'Ecological and temporal placement of early Pliocene hominids at Aramis, Ethiopia', *Nature*, vol. 371 (1994), pp. 330-1.

107 'Every feature of *Ardipithecus ramidus* seems primitive compared with other hominids . . .' The term 'primitive' has a particular meaning in cladistics and indeed in comparative biology generally. It refers to the condition of a feature with respect to the condition of the same feature in other organisms being compared: the antonym of 'primitive' in this sense is not 'advanced', but 'derived'. It is a relative term only and is not meant to represent a value judgement about crudeness or sophistication of adaptation.

108 'If *Ardipithecus ramidus* hadn't been discovered, we would have had to invent it.' See Bernard Wood's commentary accompanying the Aramis finds, *Nature*, vol. 371 (1994), pp. 280-1.

109 'Palaeontologist Elizabeth Vrba has linked this "pulse" of faunal change with changes in climate. Documenting wholesale changes in fauna demands a systematic and comprehensive approach to fossil-collecting.' Vrba's 'turnover pulse' hypothesis has proved very hard to test. Perhaps the most comprehensive attempt yet has been by A. K. Behrensmeyer and colleagues in their paper 'Late Pliocene faunal turnover in the Turkana Basin, Kenya and Ethiopia', *Science*, vol. 278 (1997), pp. 1589-94.

110 '. . . some researchers such as Bernard Wood have questioned this, suggesting that [*Homo habilis* and *H. rudolfensis*] may be no closer to modern humans than are some other extinct hominids.' See Bernard Wood and Mark Collard, 'The Human Genus', *Science*, vol. 284 (1999), pp. 65-71.

111 '*Homo erectus* . . . is arguably the earliest hominid known outside Africa.' A report by Huang Wanpo *et al.*, *Nature*, vol. 378 (1995), pp. 275-8 presents extremely tentative evidence for the presence of earlier *Homo* in China.

112 'The first fossils that can be referred to our own species . . . a recent

discovery of a skull in the Danakil Depression of Eritrea may put this back as far as a million years'. See Ernesto Abbate *et al.*, 'A one-million-year-old *Homo* cranium from the Danakil (Afar) Depression of Eritrea', *Nature*, vol. 393 (1998), 458-60.

113 '... modern Europeans could have evolved from Neandertals. Other researchers think that all modern humans descend from a relatively small population of more recent, fully modern migrants from Africa. Most evidence at present favours the latter view.' There are several excellent books on Neandertals and their relationships with modern humans. See Christopher Stringer and Clive Gamble, *In Search of the Neandertals* (London: Thames & Hudson, 1993); Christopher Stringer and Robin McKie, *African Exodus* (London: Jonathan Cape, 1996); Eric Trinkaus and Pat Shipman, *The Neandertals*, (London: Jonathan Cape, 1993), and James Shreeve, *The Neandertal Enigma*, (William Morrow, 1995).

114 'It turns out that Neandertal DNA is not only very different from that of modern Europeans but completely outside the known range of variation in *Homo sapiens* DNA.' The paper, by M. Krings and colleagues, appeared in *Cell*, vol. 90 (1997), pp. 19-30. The Neandertal bones used in this study were the original bones discovered in 1856, so they were Neandertal bones *by definition*. This forestalled any potential criticism that the researchers didn't know the provenance of the bones they were looking at.

115 'The late Allan C. Wilson of the University of California, Berkeley and his colleagues ... used the principle of parsimony to arrange these sequences into a cladogram.' See Rebecca L. Cann, Mark Stoneking and Allan C. Wilson, 'Mitochondrial DNA and human evolution', *Nature*, vol. 325 (1987), pp. 31-6.

116 'This work has since been subject to a barrage of technical criticism.' See my own piece, *Nature*, vol. 355 (1992), p. 583 for a summary of criticisms levelled at the mitochondrial DNA studies carried out by Allan Wilson and his associates.

117 'A cladogram of hominid interrelationships. Diagram by Majo Xeridat'. Redrawn with permission from Figure 38 of D. C. Johanson and B. Edgar, *From Lucy to Language* (Simon & Schuster Editions, 1996), which was taken in turn from Ian Tattersall, *The Fossil Trail* (New York: Oxford University Press, 1995).

118 '. . . *culminating in man* with his astonishing perception of the "World" around him and his powers of altering the whole fabric of the surface of large parts of the earth to suit his needs.' This passage is quoted from J. Z. Young, *The Life of Vertebrates*, 2nd edition (Oxford: Clarendon Press, 1981), p. 403.

119 '. . . palaeoanthropologists might try to use a quantitative measure such as brain volume.' A more usual measure is the 'EQ' or 'encephalisation quotient', which expresses the mass of the brain in relation to the mass of the body as a whole.

120 'These were different species from us so presumably they saw the world in different ways.' Pat Shipman and Alan Walker speculate on the world view of *Homo ergaster* in *The Wisdom of Bones* (Alfred Knopf, 1996); James Shreeve does the same thing for Neandertals in *The Neandertal Enigma* (William Morrow, 1995.)

121 'It is possible that the tools were made by *Paranthropus,* or some other unknown author, and early *Homo* never made tools at all.' This possibility was raised by Bernard Wood in a commentary, 'The oldest whodunnit in the world', *Nature,* vol. 385 (1997), pp. 292-3, on a paper by Sileshi Semaw and colleagues reporting 2.6-million-year-old stone tools from Ethiopia, *Nature,* vol. 285 (1997), pp. 333-6.

122 'A species of crow living on the island of New Caledonia . . . modifies the vanes of leaves . . . to make picks and saws for wheedling insects out of crevices . . .' See Gavin R. Hunt, 'Manufacture and use of hook-tools by New Caledonian crows', *Nature,* vol. 379 (1996), pp. 249-51.

123 'In *Trillion Year Spree*, their critical history of science fiction, Brian Aldiss and David Wingrove mark the Beast People as "kinfolk" to the Yahoos, as expressions of humanity's essentially animal nature.' Brian Aldiss and David Wingrove, *Trillion Year Spree* (Avon, 1988).

124 'In *The Language Instinct*, cognitive psychologist Stephen Pinker revives and explains Noam Chomsky's idea of a "universal grammar".' Stephen Pinker, *The Language Instinct* (Harperperennial, 1995).

125 'I refer to such things as stock-market movements and economic cycles, commonly discussed as if they were entities in their own right rather than the unpredictable results of myriad smaller, yet interacting, human processes.' In

Metaman: the Merging of Humans and Machines into a Global Superorganism (New York: Simon & Schuster, 1993), Gregory Stock sees in the increasing connectedness of human society the emergence of a single, articulate power, the summation of human activity and yet something more, something separate, a global 'superorganism'.

126 'In *Virtual History*, historian Niall Ferguson explains that for this kind of exercise to be academically valid and transcend mere entertainment, documentation already available might be used to constrain events.' Niall Ferguson (ed.), *Virtual History: Alternatives and Counterfactuals* (London: Picador, 1997).

127 'In 1993, zoologists described a hitherto unknown variety of antelope, the Vu Qiang "ox" or false oryx (*Pseudoryx nghetinhensis*), from the rugged Annamite Mountains on the Lao–Vietnamese border.' See Vu Van Dung *et al.*, 'A new species of living bovid from Vietnam', *Nature*, vol. 363 (1993), pp. 443-5.

128 'An example of this . . . is the population of Britain by the North American grey squirrel . . . at the expense of the native red squirrel.' See A. Okubo *et al.*, 'On the spatial spread of the grey squirrel in Britain', *Proceedings of the Royal Society of London*, Series B, vol. 238 (1989), pp. 113-25.

Acknowledgements

This book took much longer to write than I had imagined. My agent, Jill Grinberg at Scovil, Chichak & Galen, Inc. in New York saw it through from start to finish and deserves my first thanks. My next thanks go to her colleague Peter Robinson at Curtis Brown in London, to Stephen Morrow at Free Press, and Christopher Potter and Leo Hollis at Fourth Estate.

Many others, too numerous to mention, helped me to get to the end. In particular I should like to thank those people who took time to read and comment on all or part of the manuscript as it took shape. They were Per Erik Ahlberg, Greg Bear, Luis Chiappe, Jenny Clack, Simon Conway Morris, Peter Forey, Brian Gardiner, Rita and Tony Gee, Meave Leakey, Mark Norell, Kevin Padian, the late Colin Patterson, J. William Schopf, Neil Shubin, Bernard Wood, Tim White and Carl Zimmer.

I should also like to thank Rita and Tony Gee for providing a writer's retreat in a remote corner of France; and Gina Douglas and the staff and fellowship of the Linnean Society of London for doing much the same thing in central London.

This book would have been very much poorer without the kind invitation from Meave Leakey of the National Museums of Kenya for me to join the Museum's 1998 field season at West Turkana. I thank Meave and Louise Leakey and their colleagues and field crew for showing me what real palaeontologists get up to.

ACKNOWLEDGEMENTS

I should also like to thank my indulgent colleagues at *Nature* who filled in for me while my mind and occasionally my body were elsewhere. Joanne Webber handled a lot of tedious administration, thus making my life very much easier; and Majo Xeridat drew the pictures. I apologise if I have omitted, through oversight, the name of anyone not in this list equally deserving of appreciation. Needless to say, the mistakes and opinions in this book are mine.

The following people, companies and institutions were kind enough to permit the use of quotations, copyright material or registered trademarks: Per Erik Ahlberg, the American Association for the Advancement of Science, the American Museum of Natural History, Greg Bear, Bestfoods Ltd, Luis Chiappe, Arthur C. Clarke, Columbia University Press, Express Newspapers, Victor Gollancz, Larry Gonick, McGraw-Hill, Macmillan Magazines, Ltd, Gareth Nelson, New Directions Publishing Corporation, Oxford University Press, Penguin Books, Gerald Pollinger, Sanitarium Health Food Co., Scandinavian University Press, Solo Syndication Ltd, Caro-Beth Stewart, Mark Stoneking, Times Newspapers Ltd and the University of Chicago Press. Every effort has been made to trace and acknowledge the copyright holders for the use of copyright material in this book.

Marmite and Fred appeared as themselves. Penny was there from before the beginning until after the end. Phoebe arrived in the middle. The future belongs to her.

Index

Acanthostega 54–7, 61, 64, 67, 70, 96–7, 104, 154, 179, 181, 188

Acheulean hand axe 215

adaptation 93–9

adaptive purpose 94–7, 107, 127, 191

adaptive scenarios 117, 125, 133–4, 137, 160

African civet cat (*Civettictus civetta*) 14–15

'Age of Fishes' *see also* Devonian Period 70–1

Ahlberg, Per 60–4

AIDS 157–9

Aldiss, Brian 218

Alexander, R. McNeill 138

ALH 84001 (meteorite) 80–1

Alvarez, Walter 3, 228

alvarezsaurid 196

Alvarezsaurus 196

amino acids 155–6

ammonites 129

amniotes 35

aquatic-ape scenario 97–9, 185

Aramis collection 200, 202, 230

Archaeopteryx 176–8, 180–1, 184, 186–8, 190, 193–7

archetypes 117–20, 143

Ardipithecus ramidus 17, 200–4, 209, 226

artificial selection 33, 94–5

Asfaw, Berhane 200, 230

ATP (adenosine triphosphate) 79

Australopithecus afarensis 17, 27, 31

Australopithecus africanus 204

Australopithecus anamensis 17, 31

Australopithecus ramidus 200

bacteria 108

Bakker, Robert T. 101, 128, 178

Barbrook, Adrian C. 166

Benner, Steven 161
Benton, Mike 110–11, 129
biological species concept
    122–4, 133–4
bipedality 211–12
bird 169–71, 193–4
  evolution 171–98
  feathers 186–97
  flight 172–5, 184–6, 192–6
  flightless 197
  fossils 176–81
  hands 181–3
  relationship with dinosaurs
    178–98
'Black Skull' 15
Blanco, R. E. 78
Blaxter, Mark 154–7
Blieck, Alain 58–9, 151–2
Borges, Jorge Luis 45, 52,
    112–13, 128
brachiation 211
brain size 212–13
Brown, Frank 14–15
*Bulletin of the American Museum
    of Natural History* 139

Caenorhabditis elegans 157
calcite 71–6
Cambrian Period 67–8, 71–2
carbonates 80–1
Cartoon History of the Universe
    (Gonick) 45–6
Caudipteryx 189–91, 193
chimpanzee 202–3, 211
choanae 140–1
Chomsky, Noam 219
chordates 73–4

Clack, Jenny 54–6, 60, 64
clade 143–4
cladistics
  advantages of 151–68
  analysis of molecular
    sequence information
    156–68
  analysis of origin of birds
    179–98
  applications of 154–68
  birth of 144–5
  common ancestry and 38–44
  definition of 5–10, 39, 85–6,
    143–4, 147
  emergence of, 136–53
  evolution and 135
  flexibility of 154–68
  hypotheses and 192
  sister-group relationship 145,
    152–4, 156
cladogram 6–7, 39, 86–7, 152
  Deep Time and 43–4
  examples 39, 42–3, 62, 142,
    145–6
  fossils in 43
  outgroups 42
  using molecular information
    156–60
Coates, Mike 55–6, 64
coelacanth 48–9, 56, 64
common ancestry 34–8, 74–6,
    145, 154, 160
cladistic interpretations of
    39–44, 162–8
community of descent 79–83,
    154
comparative biology 225–6

Compsognathus 178, 188
Confuciusornis 186, 194
conodont 67–70, 77, 79
convergence 182
Cope, Edward Drinker 66
Cothurnocystis 72–7, 79
'counterfactuals' 221–7
Cradle of Life (Schopf) 81
Cretaceous Period 101, 106,
    110, 129, 180, 195
Crichton, Michael 162
Currie, Philip 189–90
Cuvier, Georges 119

Daily Express 201
Daily Mail 201
Darwin, Charles 32–3, 88–95,
    113–14, 119–21, 123–35,
    159, 176, 202
'Deep Time' 2–3, 23–7, 84–6,
    88, 147, 228
  chart of 4
  evolution as consequence of
    34
  popular perceptions of
    evolution and 3–6, 114–15
Deinonychus 101
Devonian Period 48, 51, 125–6
dinosaur
  common ancestor 163
  extinction 110–12
  feathered 189–97
  interpretations of 65–6, 101
  relationship with birds
    178–98
  renaissance 65, 101
  theropod 188–95

DNA (deoxyribose nucleic
    acid) 79, 155–6, 160–4,
    205, 209–11
  mitochondrial 206–8
Dobzhansky, Theodosius
    121–2
dromaeosaur 178–96

Echinoderm 71–6
Elginerpeton 61
Elpistostege 61, 64
elpistostegid 61–4
*Essay On the Principle of
    Population* (Malthus) 88
eurypterid 125–8, 154
Eusthenopteron 49–52, 62–4,
    103–4, 140–1, 154
evolution
  as a consequence of Deep
    Time 34
  cladist view of 135
  classical view of 117–20,
    134
  Darwinian view of 88–95,
    113–14, 133–4
  Gaskell view of 130–3
  genetics and 121–3
  influenced by multiplicity of
    factors 93–4
  linear view of 32–3, 114,
    202, 210–11, 214
  popular views of 112–14,
    126–7, 128
  progressive 133
  Romer-Simpson view of
    126–9, 133, 137, 148–53,
    175–6

evolutionary species concept 124–5, 133–4

extinction 119

false oryx (*Pseudoryx nghetinhensis*) 222

Fariña, R. A. 78

Ferguson, Niall 127, 221

fish
  abolishing archetype of 142–5
  definition of 141–5
  digits 56–8
  evolution 47–8, 56
  fins 56–62
  gills 55–6
  lobe-finned 47–51, 56, 60–4, 67, 190
  lungfishes 47–9, 56, 64, 164
  nostrils 140–2
  species 47–64

flight, evolution of 172–5, 184–98

Forey, Peter 136–40, 147, 152, 168

fossil
  imperfect interpretations of 1–2, 64, 82, 124, 141, 184–5
  labelling/processing 14–15
  search images and 67, 76, 82
  searching for 67
  sequences 5, 134
  testing hypotheses with 86

fossil record 22–3, 34, 105–9, 152–3

fossilisation, process of 27–31

Gang of Four 140–1, 148, 151–2, 168

Gardiner, Brian 136–7, 139–40, 147, 152, 168

Gaskell, W. H. 130–3, 144

gene
  discovery 121
  mutations 156
  sequencing 163–4

genetics
  advances in 155–68
  birth of 121
  relationship with linguistics 166
  synthesis with evolution 121–3

Genetics and the Origin of Species (Dobzhansky) 121

giant civet (*Pseudocivetta ingens*) 22–3, 26

Gonick, Larry 45, 57, 231

Gould, Stephen Jay 5, 7–8, 99

grade, definition of 143–4

Great Chain of Being 118–20, 128, 130, 132, 202

Great Rift, The 16–22, 27

Gulliver's Travels (Swift) 216–18

Hahn, Beatrice H. 158–9

Hallucigenia 76–7, 79

Halstead, Lambert Beverly 148–52, 168

Halsteadian Truth 150, 183

Handy Man *see Homo habilis*

Harris, John 14

Hawkes, Nigel 201

Hawkins, Benjamin
    Waterhouse 65–6, 77, 169
Hennig, Willi 6, 144–7, 168
Hesperornis 197
HIV-1 157–9
Ho, David 158–9
hominid 15
    common ancestor 207–8
    evolution 16–17, 200–27
    fossil hunters 17–23
    fossil record 15–17, 27, 31,
        199–200, 202–3, 230
    interrelationships 208–10
'Hominid Gang' 19
Homo erectus 8, 31, 204, 215,
    220, 222–3
Homo ergaster 204, 212, 215
Homo habilis 17, 19, 204, 214
Homo heidelbergensis 204
Homo rudolfensis 204
Homo sapiens
    African origin 205–8
    brain size 212–13
    common ancestor 207–9
    defining features 220
    evolution 8, 204–27
    fossil record 204–5
    solitary state 220–7
hyoid 55
hyomandibula 55
hypothesis
    definition of 86–8
    strength of 151
    untestable 95–7

Ichthyostega 51–7, 64
Iguanodon 65–6, 169

intelligence, measuring 213–14
Island of Dr Moreau, The
    (Wells) 216–18

Jarvik, Erik 49, 51, 53–5, 67
Jefferies, Richard 74–6
Johanson, Donald 27
Johanson, Zerina 62–3
Jurassic Park 111, 162–3

Kafka and His Precursors
    (Borges) 45, 52, 112–13,
    128
KNM-WT 15000 *see*
    Nariokotome Boy
KNM-WT 17000 *see* Black
    Skull
knuckle-walking 211

Lake Turkana 11–20, 28
lamprey 59–60
lancelet 73, 130
Language Instinct, The (Pinker)
    219
language, relationship with
    intelligence of 216–21
Leakey, Louis 18–19
Leakey, Louise 19
Leakey, Mary 18–19
Leakey, Meave 13–15, 19, 22,
    44
Leakey, Richard 19
Lebedev, Oleg 54
Lewontin, Richard C. 99
Liaoning fossil beds 186–90
Life of Vertebrates (Young)
    209–10

lineal ancestors 22–3
linguistics, relationship with genetics of 166–7
Linnaeus 115–20, 134
Linnean Society of London 130
LO5 14–15, 21, 23, 44, 200
Longbottom, Alison 139
Longisquama 179–81
lobopodians 77
'Lucy' 17, 27
'Lungfishes, Tetrapods, Paleontology and Plesiomorphy' (Rosen, Forey, Patterson & Gardiner) 139–40
Lyell, Charles 32, 88, 119

Madagascar Star orchid 90–3, 154
Malthus, Thomas 88, 90
Mars *Rover* 82
Marsh, Othniel Charles 66
Mayr, Ernst 122–4, 133
McKay, David S. 80
McPhee, John 2–3
Megalancosaurus 179–81
Megalosaurus 65–6, 168
Megatherium 78–9, 188
Meyer, Axel 159
'missing links' 7, 32, 44, 53, 62–4, 74–5, 111, 133–4, 201
mitochondria 206
Mitochondrial Eve 208–10
Modern Synthesis 121–3, 125, 128
Mononykus 195–6

Morris, Simon Conway 76–7
Mouland, Bill 201

Nariokotome Boy 15
Natural History Museum, 9, 46–7, 50, 52, 57–9, 65, 69, 71, 74, 136–41, 148–50, 176, 200
    Fossil Fish Section 9, 136–9
    Hall of Fossil Fishes 46–7, 57–8, 65, 71, 138
natural selection
    analogous presentation of 32–3
    artificial selection and 94–5, 114
    discovery of 88–9
    female preference 159–60
    fossil record and 125
    misinterpretations of 113–14
    principles of 133–4
    simplicity of 89–90, 119
    testable hypotheses 90–5
Nature 61, 148–9, 162, 189, 191, 199–203
nature, classical view of 117–20
Neandertal Man (*Homo neanderthalensis*) 204–5, 208, 210, 213, 223–5
Negotium perambulans (E. F. Benson) 108–9
Nelson, Gareth 147, 168
'neo-Darwinian' 121–3
Newton, Sir Isaac 87, 113
Ngeneo, Bernard 19
Nilsson, L. Anders 90, 92–3
node, definition of 36–7, 145

Norell, Mark 189–90

North American grey squirrel (*Sciurus carolinensis*) 224–5

notochord 73

Nutcracker Man *see Paranthropus boisei*

Nzube, Peter 19, 21, 44

Occam's Razor *see also* Principle of Parsimony 6–7, 40–1

On the Various Contrivances by which British and Foreign Orchids are Fertilized by Insects (Darwin) 90–2

Ordovician Period 72, 74

Origin of Species (Darwin) 32, 94, 113–14, 120–1, 128, 130, 176

osteolepiform 49–50, 58, 62–3, 141

Ostrom, John 101, 178, 187

outgroup 41

Oviraptor 192

Owen, Richard 65, 176

Ozymandias (Shelley) 11, 30–1

Pääbo, Svante 205, 208

PAHs (polyaromatic hydrocarbons) 80–1

palaeontology
nature of 8
voodoo 128–9, 132, 191, 202

pancreatic ribonuclease 160–2

Paranthropus aethiopicus 15

Paranthropus boisei 17–18, 212, 214

Parrington, Rex 142–3, 168

Patterson, Colin 9, 136–9, 142–3, 147, 149–52, 168, 176

Pei-Ji, Chen 187

Periodic Table 115–20, 128, 132

phylogenetic systematics *see* cladistics

Pinker, Stephen 219

Principle of Parsimony *see also* Occam's Razor 6–7, 40–1, 182–3

Principles of Geology, The (Lyell) 88

Proconsul africanus 18

Protarchaeopteryx 189–91, 193

proteins 155–6

pteraspid fishes 58–60, 69–71, 137–8, 151–2, 154

pterosaur 194

Qiang, Ji 189–91

racial senescence 128–9

red squirrel (*Sciurus vulgaris*) 224–5

reproductive isolation 122–4

Return of the Dancing Dinosaurs, The (Bakker) 101

rhizodont 63–4

RNA (ribose nucleic acid) 79, 160–2

Romer, Alfred Sherwood 125–9, 133, 137, 140, 143, 148–53

Rosen, Donn 8, 139, 148, 150–2, 158, 183
roundworm 105–6, 154–7
ruminant 160

Säve-Söderbergh, Gunnar 51, 54, 67
scenario 96–7
Schopf, J. William 81, 108
search image 59–60, 64, 67, 82, 84, 177–8
sexual reproduction, as evolutionary response to disease 107–9
Shelley, Percy Bysshe 11, 30–1
Shipman, Pat 215
Shu-An, Ji 189–91
Silurian Period 125–6
Simpson, George Gaylord 124, 127–9, 133–4, 137, 140, 143, 148–53
Sinosauropteryx 187–93
sister-group relationship 145, 152–4, 156, 166, 180
sloth 78–9
Society of Vertebrate Paleontology 187
spandrels 99–101
species, definition of 122
Stewart, Caro–Beth 162
Swift, Jonathan 216–18
swordtail 159–60
Systema Naturae 115

taxonomy 115–20, 151–2
tetrapods 35, 47–50, 67, 141–2, 190

Carboniferous 51–2
Devonian 54
digits 53–64, 106, 171
limbs 85–6
relationship with other lobe-finned fishes 60–4
skin covering 188–96
theory of relativity 87
theory, definition of 87
therizinosaur 189
theropod 178, 188–95
time, perception of 23–6
tools, ability to use 214–15
topology, definition of 37–9
Triassic Period 48, 68
Triceratops 101–3, 124, 128, 153
cladogram 104, 105–7
Trillion Year Spree (Aldiss & Wingrove) 218
triplet code 155–6

uncounted species 30
Unenlagia 192

Velociraptor 48, 178, 180, 190, 192, 197
velvet worms 77, 79
vertebrate
anatomy 68–70
evolution 71, 184, 209–10
origin of 130–3
Viking lander 82
Virtual History (Ferguson) 221
viruses, evolution of 157–9
von Linné, Carl see Linnaeus
Vu Qiang 'ox' see false oryx 222

Walker, Alan 215, 230
Wallace, Alfred Russel 90
Walton, Ashley 201
Wells, H. G. 216–18
Westoll, Stanley 61
Weston, Eleanor 13, 15
White, Tim 199–201
Wilson, Allan C. 205, 207
Wingrove, David 218
Wisdom Of Bones (Walker &
    Shipman) 215
WoldeGabriel, Giday 201
Wonderful Life (Gould) 7, 228
Wood, Bernard 202, 204

Young, J. Z. 209–10
Young, Sally 139

Zinj 17 *see also Paranthropus
    boisei*